Attacking Problems in Logarithms and Exponential Functions

David S. Kahn

DOVER PUBLICATIONS
Garden City, New York

Bibliographical Note

Attacking Problems in Logarithms and Exponential Functions is a new work,
first published by Dover Publications in 2015.

Library of Congress Cataloging-in-Publication Data

Kahn, David S.
Attacking problems in logarithms and exponential functions / David S. Kahn.
    p. cm.
ISBN-13: 978-0-486-79346-7
ISBN-10: 0-486-79346-X
1. Exponents (Algebra)—Textbooks. 2. Logarithmic functions—Textbooks. I. Title.
QA161.E95E97 2014
512.9'22076—dc23

2015021790

Printed in Canada by Marquis Book
79346X07   2024
www.doverpublications.com

To the Reader:

Logarithms and exponential functions are topics that students usually encounter in Precalculus, or sometimes in Advanced Algebra. They are of crucial importance to understand both higher mathematics and all of the sciences, as well as Economics, Finance, and a host of other topics. Fortunately, they are not that difficult to master, with a little diligence and patience. Thanks to the book that you are about to use, you will be able to attack logarithms and exponential functions and successfully conquer them.

One of the keys to doing well in these areas is to be comfortable with graphing and with algebraic manipulation. Logarithms and exponential functions require an understanding of how a function grows, and this will be easy for you to understand when you picture the graph. You will also learn to solve equations with logarithms or containing exponentials, and the algebra is not that hard. Sometimes we will show you all of the steps to a solution; other times, we will leave you to work them out on your own. If you are not already good at factoring a quadratic equation, or working with distributing and rearranging terms, we suggest that you practice those topics. It will make this book much easier to use!

Another key to doing well is to drill. It is important to be able to work with logarithms, which is not nearly as scary as it might sound, and it will be much easier if you do all of the practice problems to reinforce the concepts. If you do so, you will find that logarithms and exponential functions are not so hard after all.

We have organized this book so that you can proceed from one topic to another, or you can jump to the topics that you want to work on. The book is divided into ten units, each of which will teach you what you need to know to do well in that topic. This is not designed to be an exhaustive treatise on logarithms or exponential functions, nor is it designed to be a textbook. Rather, this book focuses on the essentials and how to get the problems right. We suggest that you read through each unit completely and do all of the exercises and practice problems. Each example and problem has a complete explanation to help you understand how to solve the problem correctly. After you have worked through a unit, if you still want more practice problems, you may want to go to a textbook.

Logarithms and exponential functions are a highly valuable and useful area of mathematics. After you have gone through this book, you will be able to handle with ease problems involving logarithms and exponents on your exams, and apply this knowledge to mathematics, the sciences, the social sciences, and business. Are you ready? Then it's time to Attack Logarithms and Exponential Functions!

Acknowledgments

First of all, I would like to thank Nicole Maisonet for her excellent drawings. It is a tedious task and she did it with grace and enthusiasm. Next, I would like to thank Magan Farraj for working through all of the problems, double-checking my calculations. I owe a lifetime debt to my father, Peter Kahn, and to my dear friend, Arnold Feingold, who encouraged my interest in Mathematics and have always been there to guide me through the rough spots. And finally, I would like to thank the very many students whom I have taught and tutored, who never hesitate to correct me when I am wrong, and who provide the fulfillment that I so deeply derive from teaching math.

Table of Contents

Attacking Problems in Logarithms and Exponential Functions

UNIT ONE

Seven Simple Rules for Working with Exponents

When we multiply a number by itself, we call this process *squaring*. For example, when we multiply $3 \cdot 3 = 9$ we use a short-hand method. We write the expression as 3^2, where the 2 indicates that we have multiplied the number by itself 2 times. So, if we wrote 3^3, that would indicate that we have multiplied the number by itself 3 times, that is, $3 \cdot 3 \cdot 3 = 27$. Similarly, $3^4 = 3 \cdot 3 \cdot 3 \cdot 3 = 81$, and so on. Thus, $5^6 = 5 \cdot 5 \cdot 5 \cdot 5 \cdot 5 \cdot 5 = 15,625$. Got the idea?

Notice how quickly the result grows when we multiply a number by itself. For example, we only had to multiply 5 by itself 6 times to get 15,625. Think about what this means. If we had 5 people in a room, and each of them brought 5 friends into the room, we would have $5^2 = 25$ people. If we now had each of those people bring in 5 friends, we would have $5^3 = 125$ people. If we then repeated the process three more times, we would have 15,625 people. We would need a very big room! This is the essential fact of exponential growth, which we will explore in this book. Namely, that one can start with a small number and, in just a few steps, end up with a very large number.

Now let's figure out some rules about exponents. What does it mean to raise a number to the power of 1? Let's see. If we write 4^1, we multiply 4 one time. That is, $4^1 = 4$. Now let's dispense with using numbers. If we write x^1, we multiply x, one time, so $x^1 = x$.

Now, what if we raise x to the power a. This means that we multiply x by itself a times.

$$\text{This is our first rule: } x^a = \underbrace{x \cdot x \cdot x ... x}_{a}.$$

Now suppose we have two expressions, x^3 and x^5. What happens when we multiply them together? Remember that $x^3 = x \cdot x \cdot x$ and $x^5 = x \cdot x \cdot x \cdot x \cdot x$, so if we multiply them, we get $x^3 \cdot x^5 = \underbrace{x \cdot x \cdot x} \cdot \underbrace{x \cdot x \cdot x \cdot x \cdot x}$. We have x multiplied by itself 8 times, so $x^3 \cdot x^5 = x^8$. Now let's do this again, but this time with x^a and x^b. If we multiply them, we get $x^a \cdot x^b = \underbrace{x \cdot x \cdot x ... x}_{a} \underbrace{x \cdot x \cdot x ... x}_{b}$. Thus, $x^a \cdot x^b = \underbrace{x \cdot x \cdot x ... x}_{a+b} = x^{a+b}$.

$$\text{This is our second rule: } x^a \cdot x^b = x^{a+b}.$$

Notice that this requires that the two expressions have the *same base*, namely x. What if you have $x^a \cdot y^b$? You can't do anything with this expression. For example, if you have $5^3 \cdot 4^2$, you have to evaluate each separately to get $5^3 = 125$ and $4^2 = 16$, and thus $5^3 \cdot 4^2 = 125 \cdot 16 = 2000$.

By the way, if you have $5^2 \cdot 4^2$ instead, you could evaluate them separately: $5^2 = 25$ and $4^2 = 16$. So $5^2 \cdot 4^2 = 25 \cdot 16 = 400$, or you could combine them into $(5 \cdot 4)^2 = 20^2 = 400$. This means that $x^a \cdot y^a = (xy)^a$. Be careful when you combine expressions with different bases!

Suppose that, instead of multiplying x^5 and x^3, we divide them. We get $\dfrac{x^5}{x^3} = \dfrac{\overbrace{x \cdot x \cdot x \cdot x \cdot x}^{5}}{\underbrace{x \cdot x \cdot x}_{3}}$. The x's in the numerator and denominator will cancel, leaving

us with $\dfrac{x^5}{x^3} = x^{5-3} = x^2$. Now let's do this again but with x^a and x^b. We get $\dfrac{x^a}{x^b} = \dfrac{\overbrace{x \cdot x \cdot x \ldots x}^{a}}{\underbrace{x \cdot x \cdot x \ldots x}_{b}}$. The x's in the numerator and denominator will cancel, leaving us

with $\dfrac{x^a}{x^b} = x^{a-b}$. (Note that x cannot be zero.)

> This is our third rule: $\dfrac{x^a}{x^b} = x^{a-b}$.

What if a and b are the same number? Let's divide x^5 and x^5. We get $\dfrac{x^5}{x^5} = \dfrac{\overbrace{x \cdot x \cdot x \cdot x \cdot x}^{5}}{\underbrace{x \cdot x \cdot x \cdot x \cdot x}_{5}}$. This time, all of the x's in the numerator and denominator will cancel, leaving us with $\dfrac{x^5}{x^5} = x^{5-5} = x^0 = 1$. Now let's do this again but with

x^a and x^a. We get $\dfrac{x^a}{x^a} = \dfrac{\overbrace{x \cdot x \cdot x \ldots x}^{a}}{\underbrace{x \cdot x \cdot x \ldots x}_{a}}$. The x's in the numerator and denominator will

cancel, leaving us with $\dfrac{x^a}{x^a} = x^{a-a} = x^0 = 1$.

This is not a rule, but remember $x^0 = 1$!

What does it mean to raise an exponent to an exponent? That is, what does it mean if we have an expression like $\left(x^2\right)^3$? This means that we have $\left(x^2\right)^3 = x^2 \cdot x^2 \cdot x^2$. Well, using our rule for multiplying exponential expressions, we add the exponents and get $\left(x^2\right)^3 = x^2 \cdot x^2 \cdot x^2 = x^6$. Let's do this again, but with x^a. We get $\left(x^a\right)^b = \underbrace{x^a \cdot x^a \cdot x^a \ldots x^a}_{b}$. If we add up the x^a terms, we get $\left(x^a\right)^b = \underbrace{x^a \cdot x^a \cdot x^a \ldots x^a}_{b} = x^{ab}$.

This is our fourth rule: $\left(x^a\right)^b = x^{ab}$.

It is important that you don't confuse multiplying terms where you add the exponents, with raising a term to a power, where you multiply the exponents.

What does it mean if we raise x to a negative exponent? Suppose we have x^{-4}. We can think of this as x^{0-4}. From our rule above, we know that this means $x^{-4} = x^{0-4} = \dfrac{x^0}{x^4}$. But we also know that $x^0 = 1$, so $x^{-4} = x^{0-4} = \dfrac{x^0}{x^4} = \dfrac{1}{x^4}$.

Let's do this again, but with x^a. We get $x^{-a} = x^{0-a} = \dfrac{x^0}{x^a} = \dfrac{1}{x^a}$.

This is our fifth rule: $x^{-a} = \dfrac{1}{x^a}$.

So far, we have seen what happens when we raise x to a positive number, a negative number and zero. What's left? Let's see what happens when we raise x to a fraction.

Suppose we want to evaluate 5^3. We know that $5^3 = 5 \cdot 5 \cdot 5 = 125$. If instead, we wanted to find a single power that we could raise 125 to in order to get 5, what could we do? We raised 5 to the power of 3 in order to get 125, so let's raise 125 to the $\dfrac{1}{3}$ power. This gives us: $125^{\frac{1}{3}} = \left(5^3\right)^{\frac{1}{3}}$, and using our rules of exponents, we get $125^{\frac{1}{3}} = 5^{3 \cdot \frac{1}{3}} = 5^1 = 5$. Notice that this is the cube root! In other words, $125^{\frac{1}{3}} = \sqrt[3]{125} = 5$.

Now let's do this again, but with x^a. Suppose we have $y^a = \underbrace{y \cdot y \cdot y \ldots y}_{a}$. Let's call this the result of raising $y^a = x$. Then, if we raise both sides to the power $\dfrac{1}{a}$, we get $\left(y^a\right)^{\frac{1}{a}} = x^{\frac{1}{a}}$. Using our rules of exponents, we get $\left(y^a\right)^{\frac{1}{a}} = y^{a \cdot \frac{1}{a}} = y^1 = x^{\frac{1}{a}}$. Also, remember that we got x by finding $y^a = x$, so $y = \sqrt[a]{x}$. Thus, $x^{\frac{1}{a}} = \sqrt[a]{x}$.

This is our sixth rule: $x^{\frac{1}{a}} = \sqrt[a]{x}$.

What if we see x raised to a power like $\dfrac{5}{3}$, such as with $8^{\frac{5}{3}}$? Simple. We can think of this as $\left(8^{\frac{1}{3}}\right)^5$. First we find $8^{\frac{1}{3}}$, which we now know is $\sqrt[3]{8} = 2$. Then we can evaluate

$2^5 = 32$. In other words, $8^{\frac{5}{3}} = \left(\sqrt[3]{8}\right)^5$. Of course, we could first have found $\left(8^5\right)^{\frac{1}{3}}$. But who would want to find $\sqrt[3]{32768}$?!

Now let's do this again, but with $x^{\frac{b}{a}}$. We can think of this as $\left(x^{\frac{1}{a}}\right)^b$. We know from above that $x^{\frac{1}{a}} = \sqrt[a]{x}$, so we rewrite this as $\left(\sqrt[a]{x}\right)^b$ or $\sqrt[a]{x^b}$.

> This gives us our seventh rule: $x^{\frac{b}{a}} = \left(\sqrt[a]{x}\right)^b = \sqrt[a]{x^b}$.

Here are our rules again:

Rule Number One: $x^a = \underbrace{x \cdot x \cdot x ... x}_{a}$

Rule Number Two: $x^a \cdot x^b = x^{a+b}$

Rule Number Three: $\dfrac{x^a}{x^b} = x^{a-b}$

Rule Number Four: $\left(x^a\right)^b = x^{ab}$

Rule Number Five: $x^{-a} = \dfrac{1}{x^a}$

Rule Number Six: $x^{\frac{1}{a}} = \sqrt[a]{x}$

Rule Number Seven: $x^{\frac{b}{a}} = \left(\sqrt[a]{x}\right)^b = \sqrt[a]{x^b}$

And remember that $x^0 = 1$ and $x^1 = x$!

Let's do some practice problems.

Practice Problem Set #1

Evaluate the following:

1 a) $6^1 =$ b) $6^2 =$ c) $6^0 =$ d) $6^4 =$

2 a) $4^3 \cdot 4^2 =$ b) $\dfrac{4^3}{4^2} =$ c) $\left(4^3\right)^2 =$ d) $\left(4^2\right)^3 =$

3 a) $x^5 \cdot x^9 =$ b) $y^3 \cdot y^4 =$ c) $x \cdot x^5 =$ d) $x^3 \cdot x^5 \cdot x^{10} =$

4 a) $a^{3b} \cdot a^{4b} =$ b) $y^{a+1} \cdot y^{a-1} =$ c) $z^{1+a} \cdot z^{1-a} =$ d) $x^6 \cdot x^{-6} =$

5 a) $\dfrac{x^8}{x^3} =$ b) $\dfrac{x^5}{x^{-5}} =$ c) $\dfrac{y^3}{y^5} =$ d) $\dfrac{z^2}{z^{-4}} =$

6 a) $\left(x^2\right)^7 =$ b) $\left(y^4\right)^{-2} =$ c) $\left(z^{-5}\right)^0 =$ d) $\left(a^8\right)^{\frac{1}{2}} =$

7 a) $\sqrt{y} =$ b) $\sqrt[4]{a} =$ c) $\sqrt[7]{x^3} =$ d) $\sqrt[3]{z^{-4}} =$

Solutions to Practice Problem Set #1

Evaluate the following:

1 a) $6^1 =$

Any number raised to the power 1 is itself, so $6^1 = 6$.

b) $6^2 =$

Remember from Rule Number One that 6^2 means $6 \cdot 6$, so $6^2 = 6 \cdot 6 = 36$.

c) $6^0 =$

Any number raised to the power 0 (except 0 itself) is 1, so $6^0 = 1$.

d) $6^4 =$

Remember from Rule Number One that 6^4 means $6 \cdot 6 \cdot 6 \cdot 6$, so $6^4 = 6 \cdot 6 \cdot 6 \cdot 6 = 1296$.

2 a) $4^3 \cdot 4^2 =$

Rule Number Two says that when we multiply two numbers with the same bases, we add the exponents. Here we get: $4^3 \cdot 4^2 = 4^{3+2} = 4^5 = 1024$.

b) $\dfrac{4^3}{4^2} =$

Rule Number Three says that when we divide two numbers with the same bases, we subtract the exponents. Here we get: $\dfrac{4^3}{4^2} = 4^{3-2} = 4^1 = 4$.

c) $\left(4^3\right)^2 =$

Rule Number Four says that when a number raised to a power is raised to a power, we multiply the powers. Here we get: $\left(4^3\right)^2 = 4^6 = 4096$.

d) $\left(4^2\right)^3 =$

Rule Number Four says that when a number raised to a power is raised to a power, we multiply the powers. We multiply the powers and get: $\left(4^2\right)^3 = 4^6 = 4096$. Notice that this is the same answer as problem 2c.

3   a) $x^5 \cdot x^9 =$

Rule Number Two says that when we multiply two numbers with the same bases, we add the exponents. Here we get: $x^5 \cdot x^9 = x^{5+9} = x^{14}$.

b) $y^3 \cdot y^4 =$

Rule Number Two says that when we multiply two numbers with the same bases, we add the exponents. Here we get: $y^3 \cdot y^4 = y^{3+4} = y^7$.

c) $x \cdot x^5 =$

Rule Number Two says that when we multiply two numbers with the same bases, we add the exponents. Here we get: $x \cdot x^5 = x^{1+5} = x^6$.

d) $x^3 \cdot x^5 \cdot x^{10} =$

Rule Number Two says that when we multiply two numbers with the same bases, we add the exponents. Here we get: $x^3 \cdot x^5 \cdot x^{10} = x^{3+5+10} = x^{18}$.

4   a) $a^{3b} \cdot a^{4b} =$

Rule Number Two says that when we multiply two numbers with the same bases, we add the exponents. Here we get: $a^{3b} \cdot a^{4b} = a^{3b+4b} = a^{7b}$.

b) $y^{a+1} \cdot y^{a-1} =$

Rule Number Two says that when we multiply two numbers with the same bases, we add the exponents. Here we get: $y^{a+1} \cdot y^{a-1} = y^{a+1+a-1} = y^{2a}$.

c) $z^{1+a} \cdot z^{1-a} =$

Rule Number Two says that when we multiply two numbers with the same bases, we add the exponents. Here we get: $z^{1+a} \cdot z^{1-a} = z^{1+a+1-a} = z^2$.

d) $x^6 \cdot x^{-6} =$

Rule Number Two says that when we multiply two numbers with the same bases, we add the exponents. Here we get: $x^6 \cdot x^{-6} = x^{6+(-6)} = x^0 = 1$.

5   a) $\dfrac{x^8}{x^3} =$

Rule Number Three says that when we divide two numbers with the same bases, we subtract the exponents. Here we get: $\dfrac{x^8}{x^3} = x^{8-3} = x^5$.

b) $\dfrac{x^5}{x^{-5}} =$

Rule Number Three says that when we divide two numbers with the same bases, we subtract the exponents. Here we get: $\dfrac{x^5}{x^{-5}} = x^{5-(-5)} = x^{10}$.

c) $\dfrac{y^3}{y^5} =$

Rule Number Three says that when we divide two numbers with the same bases, we subtract the exponents. Here we get: $\dfrac{y^3}{y^5} = y^{3-5} = y^{-2}$. We can use Rule Number Five to rewrite the answer as $y^{-2} = \dfrac{1}{y^2}$, if we wish. You will find that sometimes you prefer the answer in one form, and sometimes in the other form. Both are correct.

d) $\dfrac{z^2}{z^{-4}} =$

Rule Number Three says that when we divide two numbers with the same bases, we subtract the exponents. Here we get: $\dfrac{z^2}{z^{-4}} = z^{2-(-4)} = z^6$.

6 a) $\left(x^2\right)^7 =$

Rule Number Four says that when a number raised to a power is raised to a power, we multiply the powers. Here we get: $\left(x^2\right)^7 = x^{2\cdot7} = x^{14}$.

b) $\left(y^4\right)^{-2} =$

Rule Number Four says that when a number raised to a power is raised to a power, we multiply the powers. Here we get: $\left(y^4\right)^{-2} = y^{4\cdot(-2)} = y^{-8}$. We can use Rule Number Five to rewrite the answer as $y^{-8} = \dfrac{1}{y^8}$, if we wish. You will find that sometimes you prefer the answer in one form; sometimes in the other. Both are correct.

c) $\left(z^{-5}\right)^0 =$

Any number raised to the power 0 (except 0 itself) is 1, so $\left(z^{-5}\right)^0 = 1$.

d) $\left(a^8\right)^{\frac{1}{2}} =$

Rule Number Four says that when a number raised to a power is raised to a power, we multiply the powers. Here we get: $\left(a^8\right)^{\frac{1}{2}} = a^{8\cdot\frac{1}{2}} = a^4$.

7 a) $\sqrt{y} =$

Rule Number Six says that $x^{\frac{1}{a}} = \sqrt[a]{x}$, so here we get: $\sqrt{y} = y^{\frac{1}{2}}$.

b) $\sqrt[4]{a} =$

Rule Number Six says that $x^{\frac{1}{a}} = \sqrt[a]{x}$, so here we get: $\sqrt[4]{a} = a^{\frac{1}{4}}$.

c) $\sqrt[7]{x^3} =$

Rule Number Seven says that $x^{\frac{b}{a}} = \sqrt[a]{x^b}$, so here we get: $\sqrt[7]{x^3} = x^{\frac{3}{7}}$.

d) $\sqrt[3]{z^{-4}} =$

Rule Number Seven says that $x^{\frac{b}{a}} = \sqrt[a]{x^b}$, so here we get: $\sqrt[3]{z^{-4}} = z^{-\frac{4}{3}}$. We can use Rule Number Five to rewrite the answer as $z^{-\frac{4}{3}} = \dfrac{1}{z^{\frac{4}{3}}}$, if we wish.

You will find that sometimes you prefer the answer in one form, and sometimes in the other form. Both are correct.

UNIT TWO

Exponential Expressions

Now that we have learned about exponents, let's work with some more complicated problems. Often, we are faced with an expression that contains multiple exponentiated terms and we will need to simplify the expression. Usually this is because the expression is part of an equation, and we wish to simplify the expression as much as possible before we try to solve the equation. We can now apply the rules of exponents that we learned in Unit One to do the simplifying.

Example 1: Simplify the expression: $\dfrac{x^5 \cdot x^3}{x^{-4}}$.

First, let's simplify the numerator. When we multiply two exponentials with the same base, we add the exponents. Here we get $x^5 \cdot x^3 = x^{5+3} = x^8$. Now our expression is $\dfrac{x^8}{x^{-4}}$. Remember that when we divide two exponentials with the same base, we subtract the exponents. We get $\dfrac{x^8}{x^{-4}} = x^{8-(-4)} = x^{12}$.

Let's do another.

Example 2: Simplify the expression: $\dfrac{x^6 \cdot x^{-5}}{x^3 \cdot x^{-7}}$.

First, let's simplify the numerator. We add the two exponents to get $x^6 \cdot x^{-5} = x^{6+(-5)} = x^1$. Next, let's simplify the denominator. We add the two exponents to get: $x^3 \cdot x^{-7} = x^{3+(-7)} = x^{-4}$. Now our expression is $\dfrac{x^1}{x^{-4}}$. Next, we subtract the exponents to get $\dfrac{x^1}{x^{-4}} = x^{1-(-4)} = x^5$.

Let's do a more complicated one.

Example 3: Simplify the expression: $\left(x^3\right)^4 \left(x^5\right)^{-2}$.

Remember that when we raise an exponential to a power, we multiply the exponents. This gives us: $\left(x^3\right)^4 = x^{3\cdot4} = x^{12}$ and $\left(x^5\right)^{-2} = x^{5\cdot-2} = x^{-10}$. Our expression is now $x^{12} \cdot x^{-10}$. Next, we add the powers to get: $x^{12} \cdot x^{-10} = x^{12+(-10)} = x^2$.

Notice we apply the rules of exponents, one at a time, to simplify each example. Let's do one more.

Example 4: Simplify the expression: $\dfrac{\left(x^5\right)^3 \left(x^{-4}\right)^4}{\left(x^6\right)^2 \left(x^3\right)^{-7}}$.

The best way to tackle one of these is to first simplify each of the four terms individually, then combine them. In each case, we multiply the powers. In the numerator, we get: $\left(x^5\right)^3 = x^{5\cdot3} = x^{15}$ and $\left(x^{-4}\right)^4 = x^{4\cdot-4} = x^{-16}$. In the denominator, we get: $\left(x^6\right)^2 = x^{6\cdot2} = x^{12}$ and $\left(x^3\right)^{-7} = x^{3\cdot(-7)} = x^{-21}$. Now our expression is $\dfrac{x^{15}\cdot x^{-16}}{x^{12}\cdot x^{-21}}$. Then simplify the numerator by adding the exponents: $x^{15}\cdot x^{-16} = x^{15+(-16)} = x^{-1}$. Let's simplify the denominator by adding the exponents: $x^{12}\cdot x^{-21} = x^{12+(-21)} = x^{-9}$. Now our expression is $\dfrac{x^{-1}}{x^{-9}}$. We can subtract the exponents and get $\dfrac{x^{-1}}{x^{-9}} = x^{-1-(-9)} = x^8$.

In each of these examples, we simplified an expression that had several exponentiated terms, but they all had the same base. What do we do if we have terms with different bases? Simple. We work with them separately. Let's do an example.

Example 5: Simplify the expression: $\dfrac{x^5 y^3 z^{-2}}{x^{-4} y^6 z^{-4}}$.

Remember that when you divide two terms with the same base, you subtract the powers. Let's work with each term separately: $\dfrac{x^5}{x^{-4}} = x^{5-(-4)} = x^9$, $\dfrac{y^3}{y^6} = y^{3-6} = y^{-3} = \dfrac{1}{y^3}$, and $\dfrac{z^{-2}}{z^{-4}} = z^{-2-(-4)} = z^2$.

Now we can combine these and get: $x^9\cdot\dfrac{1}{y^3}\cdot z^2 = \dfrac{x^9 z^2}{y^3}$.

This was a little clumsy. There is a better way to simplify these kinds of complex, mixed expressions. Remember that $x^{-a} = \dfrac{1}{x^a}$. Similarly, $\dfrac{1}{x^{-a}} = x^a$. This means that when we have an expression with terms with negative powers in the numerator and/or denominator, we can move the term from the numerator to the denominator, or vice versa. In other words, anytime we move an exponent from the numerator to the denominator, or from the denominator to the numerator, we change its sign. So, in the previous example, we could have rewritten the expression first as $\dfrac{x^5 y^3 z^{-2}}{x^{-4} y^6 z^{-4}} = \dfrac{x^5 x^4 y^3 z^4}{y^6 z^2}$. Now we can simplify the x terms: $x^5 x^4 = x^{5+4} = x^9$ and get $\dfrac{x^9 y^3 z^4}{y^6 z^2}$. Then, we can cancel like terms and get $\dfrac{x^9 y^3 z^4}{y^6 z^2} = \dfrac{x^9 z^2}{y^3}$. On this particular problem, the second method doesn't seem to save a lot of time, but let's try it on a more complicated expression.

Example 6: Simplify the expression: $\dfrac{\left(x^5\right)^{-3}\left(y^4\right)^2\left(z^6\right)^{-3}}{\left(x^2\right)^{-4}\left(y^3\right)^5\left(z^{-2}\right)^8}$.

First, remember that when we raise an exponential to a power, we multiply the exponents. This gives us: $\dfrac{x^{-15} y^8 z^{-18}}{x^{-8} y^{15} z^{-16}}$. Now, take each term with a negative exponent and move it from the numerator to the denominator, or vice versa, and change the sign of the exponent. We now get: $\dfrac{x^{-15}y^8z^{-18}}{x^{-8}y^{15}z^{-16}} = \dfrac{x^8y^8z^{16}}{x^{15}y^{15}z^{18}}$. Now we can reduce the like terms: $\dfrac{x^8y^8z^{16}}{x^{15}y^{15}z^{18}} = \dfrac{1}{x^7y^7z^2}$.

We will sometimes encounter a problem that involves multiplying or dividing two exponential expressions. The rules are the same as before. We will simplify each of the expressions separately, then combine them and simplify again. Let's do an example.

Example 7: Simplify the expression: $\dfrac{a^3b^{-5}}{a^{-7}b^{-2}} \cdot \dfrac{a^{-8}b^4}{a^2b^{-5}}$.

Let's first rewrite the left-hand expression using positive exponents. We get: $\dfrac{a^3a^7b^2}{b^5}$. Now we combine the exponents of the like terms: $\dfrac{a^{10}}{b^3}$. Next, let's rewrite the right-hand expression using positive exponents. We get: $\dfrac{b^4b^5}{a^2a^8}$. Now we combine the exponents of the like terms: $\dfrac{b^9}{a^{10}}$. Finally, we combine our two simplified expressions: $\dfrac{a^{10}}{b^3} \cdot \dfrac{b^9}{a^{10}}$. Combining the exponents of the like terms, we get: b^6.

What if we are asked to divide the two rational expressions, instead of multiplying them? Simple! We flip the right-hand expression and multiply (just as we do with regular fractions). Then we can proceed as before. Let's do an example.

Example 8: Simplify the expression: $\dfrac{x^3y^4z^{-6}}{x^{-2}y^{10}z^{-2}} \div \dfrac{x^2y^{-10}z^{-4}}{x^{-5}y^{-8}z^3}$.

Let's first flip the right-hand term so that this becomes a multiplication problem. We get: $\dfrac{x^3y^4z^{-6}}{x^{-2}y^{10}z^{-2}} \cdot \dfrac{x^{-5}y^{-8}z^3}{x^2y^{-10}z^{-4}}$. Then rewrite the left-hand expression using positive exponents. We get: $\dfrac{x^3x^2y^4z^2}{y^{10}z^6}$. Now we can combine the exponents of the like terms: $\dfrac{x^5}{y^6z^4}$. Next, let's rewrite the right-hand expression using positive exponents. We get: $\dfrac{y^{10}z^3z^4}{x^2x^5y^8}$. Now we can combine the exponents of the like terms: $\dfrac{y^2z^7}{x^7}$. Finally, we can combine our two simplified expressions: $\dfrac{x^5}{y^6z^4} \cdot \dfrac{y^2z^7}{x^7}$. Combining the exponents of the like terms, we get: $\dfrac{z^3}{x^2y^4}$.

Sometimes we will be asked to simplify an expression involving radicals. The strategy is to simplify the expression by moving as many terms as possible out of the radical and then to reduce what is left inside the radical. Let's do an example.

Example 9: Simplify the expression: $\sqrt{x^5 y^3}$.

First, remember that the radical means "square root," which is the same as raising a term to the $\frac{1}{2}$ power. So, if we have a term to an even power, we can take the square root of that term by dividing the power by 2. For example, $\sqrt{x^6} = x^3$. If we think of x^5 as $x^4 \cdot x$ and y^3 as $y^2 \cdot y$, then we could rewrite the expression as $\sqrt{x^4 \cdot x \cdot y^2 \cdot y}$. Why would we want to do this? Because we then could take the square root of the even power terms and move them outside of the radical; in other words, $\sqrt{x^4 \cdot x \cdot y^2 \cdot y} = x^2 y \sqrt{xy}$.

Let's do another example.

Example 10: Simplify the expression: $\sqrt{x^7 y^9 z^6}$.

If we think of x^7 as $x^6 \cdot x$ and y^9 as $y^8 \cdot y$, then we could rewrite the expression as $\sqrt{x^6 x y^8 y z^6}$. Notice that we did not need to break up the z term because it is already raised to an even power. Now we can take the square root of the even power terms and move them outside of the radical: $\sqrt{x^6 x y^8 y z^6} = x^3 y^4 z^3 \sqrt{xy}$.

What if we have a cube root? It's the same concept, but we will want to divide powers by 3 instead of 2, so we will want to break up each term into powers of 3. Let's do an example.

Example 11: Simplify the expression: $\sqrt[3]{x^8 y^5 z^2}$.

If we think of x^8 as $x^6 \cdot x^2$ and y^5 as $y^3 \cdot y^2$, then we could rewrite the expression as $\sqrt[3]{x^6 x^2 y^3 y^2 z^2}$. Notice that we did not need to break up the z term because it is raised to a power less than 3. So we leave it alone. Now we take the cube root of the terms with powers that are multiples of 3 and move them outside of the radical: $\sqrt[3]{x^6 x^2 y^3 y^2 z^2} = x^2 y \sqrt[3]{x^2 y^2 z^2}$.

Of course, we can use a similar technique for a fourth root, a fifth root, and so on. Let's do another example.

Example 12: Simplify the expression: $\sqrt[5]{x^{17} y^9 z^{20}}$.

If we think of x^{17} as $x^{15} \cdot x^2$ and y^9 as $y^5 \cdot y^4$, then we could rewrite the expression as $\sqrt[5]{x^{15} x^2 y^5 y^4 z^{20}}$. Notice that we did not need to break up the z term because it is raised to a multiple of 5. When we take the fifth root of z^{20}, it is simply z^4 (because $\frac{20}{5} = 4$). Now we take the fifth root of the terms with powers that are multiples of 5 and move them outside of the radical: $\sqrt[5]{x^{15} x^2 y^5 y^4 z^{20}} = x^3 yz^4 \sqrt[5]{x^2 y^4}$.

Are you ready for some practice problems?

Practice Problem Set #2

Simplify the following expressions.

1) $\dfrac{x^{11} \cdot x^{-2}}{x^7 \cdot x^{-10}}$

2) $\dfrac{x^{16} \cdot x^{-8}}{x^{19} \cdot x^{11}}$

3) $\left(x^6\right)^6 \left(x^9\right)^{-4}$

4) $\left(x^{12}\right)^{\frac{1}{3}} \left(x^{18}\right)^{-\frac{1}{2}}$

5) $\dfrac{\left(x^8\right)^6 \left(y^{-4}\right)^5}{\left(x^3\right)^{12} \left(y^8\right)^{-2}}$

6) $\dfrac{\left(x^6\right)^{\frac{1}{3}} \left(y^{-\frac{1}{2}}\right)^4}{\left(x^8\right)^{-2} \left(y^{-10}\right)^{-\frac{1}{5}}}$

7) $\dfrac{\left(x^9\right)^{-2} \left(y^4\right)^4 \left(z^9\right)^{-1}}{\left(x^4\right)^{-4} \left(y^5\right)^3 \left(z^{-2}\right)^5}$

8) $\dfrac{\left(x^{\frac{1}{3}}\right)^3 \left(y^6\right)^{-\frac{1}{2}} \left(z^4\right)^3}{\left(x^3\right)^{-2} \left(y^3\right)^{-\frac{1}{3}} \left(z^{-2}\right)^{-2}}$

9) $\dfrac{a^4 b^{-2}}{a^7 b^{-5}} \cdot \dfrac{a^8 b^8}{a^{-7} b^3}$

10) $\dfrac{a^{-5} b^{-3}}{a^{-2} b^{-8}} \div \dfrac{a^{-3} b^6}{a^7 b^{-9}}$

11) $\dfrac{x^{10} y^8 z^{-9}}{x^{-7} y^2 z^{-5}} \div \dfrac{x^{14} y^{-4} z^{-2}}{x^{-3} y^{-10} z^2}$

12) $\sqrt{x^{11} y^8 z^7}$

13) $\sqrt{x^3 y^5 z^8}$

14) $\sqrt[3]{x^{16}y^4 z^{21}}$

15) $\sqrt[4]{x^{24}y^{10}z^5}$

Solutions to Practice Problem Set #2

Simplify the following expressions.

1) $\dfrac{x^{11} \cdot x^{-2}}{x^7 \cdot x^{-10}}$

   First, let's simplify the numerator. We add the two exponents to get $x^{11} \cdot x^{-2} = x^{11+(-2)} = x^9$.

   Next, let's simplify the denominator. We add the two exponents to get: $x^7 \cdot x^{-10} = x^{7+(-10)} = x^{-3}$.

   Now our expression is $\dfrac{x^9}{x^{-3}}$. Next, we subtract the exponents to get $\dfrac{x^9}{x^{-3}} = x^{9-(-3)} = x^{12}$.

2) $\dfrac{x^{16} \cdot x^{-8}}{x^{19} \cdot x^{11}}$

   First, let's simplify the numerator. We add the two exponents to get $x^{16} \cdot x^{-8} = x^{16+(-8)} = x^8$.

   Next, let's simplify the denominator. We add the two exponents to get: $x^{19} \cdot x^{11} = x^{19+11} = x^{30}$.

   Now our expression is $\dfrac{x^8}{x^{30}}$. Next, we subtract the exponents to get $\dfrac{x^8}{x^{30}} = x^{8-(30)} = x^{-22}$, which we can also write as $\dfrac{1}{x^{22}}$.

3) $\left(x^6\right)^6 \left(x^9\right)^{-4}$

   Remember that when we raise an exponential to a power, we multiply the exponents. This gives us: $\left(x^6\right)^6 = x^{6\cdot 6} = x^{36}$ and $\left(x^9\right)^{-4} = x^{9(-4)} = x^{-36}$. Our expression is now $x^{36} \cdot x^{-36}$. Next, we add the powers to get: $x^{36} \cdot x^{-36} = x^{36+(-36)} = x^0 = 1$.

4) $\left(x^{12}\right)^{\frac{1}{3}} \left(x^{18}\right)^{-\frac{1}{2}}$

   Remember that when we raise an exponential to a power, we multiply the exponents. This gives us: $\left(x^{12}\right)^{\frac{1}{3}} = x^{12\cdot\frac{1}{3}} = x^4$ and $\left(x^{18}\right)^{-\frac{1}{2}} = x^{18\left(-\frac{1}{2}\right)} = x^{-9}$. Our expression is now $x^4 \cdot x^{-9}$. Next, we add the powers to get: $x^4 \cdot x^{-9} = x^{4+(-9)} = x^{-5}$, which we can also write as $\dfrac{1}{x^5}$.

5) $\dfrac{\left(x^8\right)^6 \left(y^{-4}\right)^5}{\left(x^3\right)^{12} \left(y^8\right)^{-2}}$

First, remember that when we raise an exponential to a power, we multiply the exponents. This gives us: $\dfrac{x^{48} y^{-20}}{x^{36} y^{-16}}$. Now, take each term with a negative exponent and move it from the numerator to the denominator, or vice versa, and change the sign of the exponent. We now get: $\dfrac{x^{48} y^{-20}}{x^{36} y^{-16}} = \dfrac{x^{48} y^{16}}{x^{36} y^{20}}$. Now we reduce the like terms: $\dfrac{x^{48} y^{16}}{x^{36} y^{20}} = \dfrac{x^{12}}{y^4}$.

6) $\dfrac{\left(x^6\right)^{\frac{1}{3}} \left(y^{-\frac{1}{2}}\right)^4}{\left(x^8\right)^{-2} \left(y^{-10}\right)^{-\frac{1}{5}}}$

When we raise an exponential to a power, we multiply the exponents. This gives us: $\dfrac{x^2 y^{-2}}{x^{-16} y^2}$. Now, take each term with a negative exponent and move it from the numerator to the denominator, or vice versa, and change the sign of the exponent. We now get: $\dfrac{x^2 y^{-2}}{x^{-16} y^2} = \dfrac{x^2 x^{16}}{y^2 y^2}$. Now we combine the like terms: $\dfrac{x^2 x^{16}}{y^2 y^2} = \dfrac{x^{18}}{y^4}$.

7) $\dfrac{\left(x^9\right)^{-2} \left(y^4\right)^4 \left(z^9\right)^{-1}}{\left(x^4\right)^{-4} \left(y^5\right)^3 \left(z^{-2}\right)^5}$

When we raise an exponential to a power, we multiply the exponents. This gives us: $\dfrac{x^{-18} y^{16} z^{-9}}{x^{-16} y^{15} z^{-10}}$. Now, take each term with a negative exponent and move it from the numerator to the denominator, or vice versa, and change the sign of the exponent. We now get: $\dfrac{x^{-18} y^{16} z^{-9}}{x^{-16} y^{15} z^{-10}} = \dfrac{x^{16} y^{16} z^{10}}{x^{18} y^{15} z^9}$. Now we reduce the like terms: $\dfrac{x^{16} y^{16} z^{10}}{x^{18} y^{15} z^9} = \dfrac{yz}{x^2}$.

8) $\dfrac{\left(x^{\frac{1}{3}}\right)^3 \left(y^6\right)^{-\frac{1}{2}} \left(z^4\right)^3}{\left(x^3\right)^{-2} \left(y^3\right)^{-\frac{1}{3}} \left(z^{-2}\right)^{-2}}$

When we raise an exponential to a power, we multiply the exponents. This gives us: $\dfrac{x^1 y^{-3} z^{12}}{x^{-6} y^{-1} z^4}$. Now, take each term with a negative exponent and move it from the numerator to the denominator, or vice versa, and change

the sign of the exponent. We now get: $\dfrac{x^1 y^{-3} z^{12}}{x^{-6} y^{-1} z^4} = \dfrac{x^1 x^6 y^1 z^{12}}{y^3 z^4}$. Now we reduce

the like terms: $\dfrac{x^1 x^6 y^1 z^{12}}{y^3 z^4} = \dfrac{x^7 z^8}{y^2}$.

9) $\dfrac{a^4 b^{-2}}{a^7 b^{-5}} \cdot \dfrac{a^8 b^8}{a^{-7} b^3}$

Let's first rewrite the left-hand expression using positive exponents. We get:

$\dfrac{a^4 b^5}{a^7 b^2}$. Now we can combine the exponents of the like terms: $\dfrac{b^3}{a^3}$. Next, let's

rewrite the right-hand expression using positive exponents. We get: $\dfrac{a^8 a^7 b^8}{b^3}$.

Now we combine the exponents of the like terms: $a^{15} b^5$. Finally, we combine

our two simplified expressions: $\dfrac{b^3}{a^3} \cdot a^{15} b^5$. Combining the exponents of the

like terms, we get: $a^{12} b^8$.

10) $\dfrac{a^{-5} b^{-3}}{a^{-2} b^{-8}} \div \dfrac{a^{-3} b^6}{a^7 b^{-9}}$

Let's first flip the right-hand term so that this becomes a multiplication

problem. We get: $\dfrac{a^{-5} b^{-3}}{a^{-2} b^{-8}} \cdot \dfrac{a^7 b^{-9}}{a^{-3} b^6}$. Now let's rewrite the left-hand expression

using positive exponents. We get: $\dfrac{a^2 b^8}{a^5 b^3}$. Now we combine the exponents of

the like terms: $\dfrac{b^5}{a^3}$ Next, let's rewrite the right-hand expression using positive

exponents. We get: $\dfrac{a^7 a^3}{b^9 b^6}$. Now we combine the exponents of the like terms:

$\dfrac{a^{10}}{b^{15}}$. Finally, we can combine our two simplified expressions: $\dfrac{b^5}{a^3} \cdot \dfrac{a^{10}}{b^{15}}$.

Combining the exponents of the like terms, we get: $\dfrac{a^7}{b^{10}}$.

11) $\dfrac{x^{10} y^8 z^{-9}}{x^{-7} y^2 z^{-5}} \div \dfrac{x^{14} y^{-4} z^{-2}}{x^{-3} y^{-10} z^2}$

Let's first flip the right-hand term so that this becomes a multiplication

problem. We get: $\dfrac{x^{10} y^8 z^{-9}}{x^{-7} y^2 z^{-5}} \cdot \dfrac{x^{-3} y^{-10} z^2}{x^{14} y^{-4} z^{-2}}$. Now let's rewrite the left-hand

expression using positive exponents. We get: $\dfrac{x^{10} x^7 y^8 z^5}{y^2 z^9}$. Now we combine

the exponents of the like terms: $\dfrac{x^{17} y^6}{z^4}$. Next, let's rewrite the right-hand expression using positive exponents. We get: $\dfrac{y^4 z^2 z^2}{x^3 x^{14} y^{10}}$. Now we combine the exponents of the like terms: $\dfrac{z^4}{x^{17} y^6}$. Finally, we combine our two simplified expressions: $\dfrac{x^{17} y^6}{z^4} \cdot \dfrac{z^4}{x^{17} y^6}$. Combining the exponents of the like terms, we get 1:

12) $\sqrt{x^{11} y^8 z^7}$

If we think of x^{11} as $x^{10} \cdot x$ and z^7 as $z^6 \cdot z$, then we could rewrite the expression as $\sqrt{x^{10} x y^8 z^6 z}$. Notice that we did not need to break up the y term because it is already raised to an even power. Now we take the square root of the even powered terms and move them outside of the radical: $\sqrt{x^{10} x y^8 z^6 z} = x^5 y^4 z^3 \sqrt{xz}$.

13) $\sqrt{x^3 y^5 z^8}$

If we think of x^3 as $x^2 \cdot x$ and y^5 as $y^4 \cdot y$, then we could rewrite the expression as $\sqrt{x^2 x y^4 y z^8}$. Notice that we did not need to break up the z term because it is already raised to an even power. Now we take the square root of the even powered terms and move them outside of the radical: $\sqrt{x^2 x y^4 y z^8} = xy^2 z^4 \sqrt{xy}$.

14) $\sqrt[3]{x^{16} y^4 z^{21}}$

If we think of x^{16} as $x^{15} \cdot x$ and y^4 as $y^3 \cdot y$, then we could rewrite the expression as $\sqrt[3]{x^{15} x y^3 y z^{21}}$. Notice that we did not need to break up the z term because its power is a multiple of 3. So we leave it alone. Now we take the cube root of the terms with powers that are multiples of 3 and move them outside of the radical: $\sqrt[3]{x^{15} x y^3 y z^{21}} = x^5 y z^7 \sqrt[3]{xy}$.

15) $\sqrt[4]{x^{24} y^{10} z^5}$

If we think of y^{10} as $y^8 \cdot y^2$ and z^5 as $z^4 \cdot z$, then we could rewrite the expression as $\sqrt[4]{x^{24} y^8 y^2 z^4 z}$. Notice that we did not need to break up the x term because its power is a multiple of 4. So we leave it alone. Now we take the fourth root of the terms with powers that are multiples of 4 and move them outside of the radical: $\sqrt[4]{x^{24} y^8 y^2 z^4 z} = x^6 y^2 z \sqrt[4]{y^2 z}$.

UNIT THREE

Scientific Notation

One area where we often will encounter exponents is *scientific notation*. This is a convenient way to represent very large or very small numbers. Scientific notation uses powers of ten multiplied by a number, instead of writing out the whole number. Many students first encounter scientific notation in Chemistry, Physics, and other sciences, where it is generally used to represent very large or very small numbers. For example, in Chemistry, Avogadro's constant is 6.02×10^{23}, which is a lot easier to work with than 602,000,000,000,000,000,000,000! Or, suppose we were doing an Astronomy problem that involved the distance to a star that is 90,000,000,000 miles away. We probably don't want to write the number over and over again. We can write the number a different way that will be much less cumbersome using scientific notation. Notice that 90,000,000,000 has ten zeros, and is eleven digits long. Let's now look at the powers of ten to see how they will help us.

$$10^1 = 10$$
$$10^2 = 100$$
$$10^3 = 1000$$
$$10^4 = 10,000$$
and so on.

Notice that when we raise 10 to a positive integer power, we get a one, followed by zeros. In fact, the number of zeros that we get is equal to the power to which we raised 10. For example, 10 raised to the third power is 1000, which is a 1, followed by three zeros. So, if we raised 10 to the tenth power, we should get a 1, followed by ten zeros. That is, $10^{10} = 10,000,000,000$. Next, let's multiply this number by 9. We get 90,000,000,000, which is the number that we want! This means that we could think of 90,000,000,000 as multiplying 9 by 10^{10}, that is $90,000,000,000 = 9 \times 10^{10}$. Writing a number in this shorthand format is what we mean by scientific notation.

Let's do a couple of examples.

Example 1: Write 4,000,000,000 in scientific notation.

Count the number of zeros. We get 9. We know that 10^9 is 1,000,000,000, so we can think of 4,000,000,000 as 4 multiplied by 10^9. We write this in scientific notation as 4×10^9.

Example 2: Write 17,000,000,000,000 in scientific notation.

Again, we count the number of zeros. This time, we get 12. We know that 10^{12} is 1,000,000,000,000 so we can think of 17,000,000,000,000 as 17 multiplied by 10^{12}. We write this in scientific notation as 17×10^{12}.

It is customary to write numbers in scientific notation using only a single digit to the left of the decimal point. Thus, in example 2, we would write 17×10^{12} as 1.7×10^{13}. How did we get this? We can think of 17 as 1.7×10, which means that we need to add a 1 to the power in 17×10^{12}, giving us 1.7×10^{13}. Let's do a couple of examples.

Example 3: Write 430,000,000 in scientific notation.
We count the number of zeros and get 7, so we can write this as 43×10^7. Next, we can write 43 as 4.3×10. This means that we need to add 1 to the power in 43×10^7, which gives us 4.3×10^8.

Example 4: Write 517,000,000 in scientific notation.
We count the number of zeros and get 6, so we can write this as 517×10^6. Next, we can write 517 as 5.17×100. This means that we need to add 2 to the power in 517×10^6, which gives us 5.17×10^8.

Now we have a rule for how to write numbers in scientific notation.

Start at the leftmost number. Count the number of digits to the right of that first number, which will tell us the power of 10 that we will use in the scientific notation of the number.

Let's do another couple of examples.

Example 5: Write 60,200,000,000,000 in scientific notation.
Start at the leftmost number, which is 6. From there, we have 13 digits. So, we can write this as 6.02×10^{13}.

Example 6: Write 9,426,000,000 in scientific notation.
Start at the leftmost number, which is 9. From there, we have nine digits. So, we can write this as 9.426×10^9.

We can also use scientific notation to represent decimal numbers. Suppose we want to write the number 0.00005 in scientific notation. This can be written as the fraction $\dfrac{5}{100,000}$. Now, recall that $\dfrac{1}{100,000}$ is 10^{-5}. Therefore, we can write 0.00005 as 5×10^{-5}. It is cumbersome to figure out the fraction each time, so a convenient rule it to count the number of zeros (to the right of the decimal point) before the first nonzero number and add 1. The power of 10 is the negative of that number. Here, we have 4 zeros, so the power is −5.

Example 7: Write 0.00000008 in scientific notation.
We count the number of zeros and then add 1. Thus the power of 10 is −8 and we can rewrite 0.00000008 as 8×10^{-8}.

What if we have a number like 0.0000453? We count the zeros and add 1, then negate that number, which gives us −5. Therefore, 0.0000453 as 4.53×10^{-5}.

Example 8: Write 0.0000000612 in scientific notation.
We count seven zeros, so the power of 10 is −8. Thus, we can write 0.0000000612 as 6.12×10^{-8}.

If we are given a number in scientific notation and we want to convert it to ordinary decimal notation, we look at the power of 10. If it is a positive integer, we start at the decimal point and move it that number of places to the right, filling all of the spaces with either the digits from the decimal or zeros. If it is a negative integer, we move the decimal point the appropriate number of places to the left. Let's do an example.

Example 9: Convert 4.9×10^5 into ordinary decimal notation.

Here, 10 is raised to the power 5, so we need to move the decimal point five places to the right. We get: $4.9 \times 10^5 = 490,000$. Let's do another example.

Example 10: Convert 2.67×10^8 into ordinary decimal notation.

Here, 10 is raised to the power 8, so we need to move the decimal point eight places to the right. We get: $2.67 \times 10^8 = 267,000,000$.

Let's do an example with a negative power.

Example 11: Convert 3.54×10^{-6} into ordinary decimal notation.

Here, 10 is raised to the power −6, so we need to move the decimal point six places to the left. We get: $3.54 \times 10^{-6} = 0.00000354$.

What if we want to add or subtract two numbers in scientific notation?

Example 12: Find the sum of $5.3 \times 10^7 + 2.6 \times 10^7$.

We simply add 5.3 and 2.6 and multiply them by 10^7, giving us 7.9×10^7.

Of course, we could also subtract two numbers.

Example 13: Find the difference of $6.4 \times 10^5 - 5.2 \times 10^5$.

Again, we subtract 6.4 and 5.2 and multiply them by 10^5, giving us 1.2×10^5. Note what would happen if, instead, we add $6.4 \times 10^5 + 5.2 \times 10^5$. We would get 11.6×10^5. If we want to write this using our standard format, we can make this 1.16×10^6. Why? The number 11.6×10^5 means that we write 116 and four more zeros, for a total of 5 digits to the right of the decimal point, giving us 1,160,000. The number 1.16×10^6 again means that we write 116 followed by four more zeros, giving us 1,160,000. So, if we shift the decimal point one place to the left, we have to compensate by adding 1 to the power of 10.

Let's do another example.

Example 14: Find the sum of $9.3 \times 10^8 + 4.1 \times 10^8$.

We add 9.3 and 4.1, and multiply them by 10^8, giving us 13.4×10^8. We can rewrite this as 1.34×10^9. Note that we don't have to rewrite the answer; it is just customary to do so.

What if we wanted to add two numbers with different powers of 10?

Example 15: Find the sum of $7.6 \times 10^4 + 2.2 \times 10^5$.

What we do is first rewrite one of the numbers so that it has the same power of 10 as the other number. It is generally easier to do the arithmetic when we rewrite the number with the higher power. Let's rewrite 2.2×10^5 as 22×10^4. Now we add the two numbers to get $7.6 \times 10^4 + 22 \times 10^4 = 29.6 \times 10^4$, which we then rewrite as 2.96×10^5.

Now let's multiply two numbers in scientific notation.

Example 16: Find the product of $(1.2 \times 10^3)(2.5 \times 10^4)$.

What we do here is multiply 1.2 and 2.5, and add the powers of 10. Why do we add the powers? Remember your rules of exponents: $10^a \cdot 10^b = 10^{a+b}$. Here, we get $1.2 \cdot 2.5 = 3$, so $(1.2 \times 10^3)(2.5 \times 10^4) = 3 \times 10^7$.

Example 17: Find the product of $(5 \times 10^9)(9 \times 10^3)$.

We multiply 5 and 9, and we add the powers 9 and 3. We get: $(5 \times 10^9)(9 \times 10^3) = 45 \times 10^{12}$, which we rewrite as 4.5×10^{13}.

The rules are the same when we have numbers with negative powers.

Example 18: Find the product of $(6 \times 10^{-5})(4 \times 10^{-3})$.

We multiply 6 and 4, and we add the powers –5 and –3. We get: $(6 \times 10^{-5})(4 \times 10^{-3}) = 24 \times 10^{-8}$, which we rewrite as 2.4×10^{-7}.

Division follows the same rules as multiplication, except that you subtract the powers of 10.

Example 19: Find the quotient of $\dfrac{4.2 \times 10^7}{2.4 \times 10^3}$.

Dividing 4.2 by 2.4 might be bothersome, so first we can rewrite this as $\dfrac{42 \times 10^6}{24 \times 10^2}$.

Now we divide 42 by 24, and we subtract the powers 6 and 2. We get $\dfrac{42}{24} = 1.75$, and $\dfrac{10^6}{10^2} = 10^{6-2} = 10^4$, which gives us $\dfrac{42 \times 10^6}{24 \times 10^2} = 1.75 \times 10^4$.

If you practice, you will find that scientific notation seems difficult at first, but really isn't. You just have to remember the rules of where you put the decimal point.

Are you ready to practice?

Practice Problem Set #3

Practice Problem 1: Convert the following numbers to scientific notation:
a) 53,000   b) 572,000,000,000   c) 0.00068   d) 0.0000000914
e) 0.00003142

Practice Problem 2: Convert the following numbers to ordinary decimal notation:
a) 6.4×10^3 b) 3.7×10^5 c) 2.735×10^7 d) 5.4×10^{-6}
e) 9.778×10^{-9}

Practice Problem 3: Find the following sums:
a) $3.2 \times 10^4 + 5.2 \times 10^4 =$

b) $6.6 \times 10^5 + 8.2 \times 10^5 =$

c) $3.5 \times 10^8 + 4.1 \times 10^7 =$

Practice Problem 4: Find the following differences:
a) $5.2 \times 10^6 - 3.2 \times 10^6 =$

b) $4.9 \times 10^5 - 8.2 \times 10^5 =$

c) $5.1 \times 10^8 - 3.9 \times 10^7 =$

Practice Problem 5: Find the following products:
a) $\left(2.5 \times 10^3\right)\left(4 \times 10^5\right) =$

b) $\left(3.6 \times 10^6\right)\left(8 \times 10^7\right) =$

c) $\left(9 \times 10^9\right)\left(1.6 \times 10^{-19}\right) =$

Practice Problem 6: Find the following quotients:
a) $\dfrac{2.5 \times 10^9}{5 \times 10^3} =$

b) $\dfrac{4 \times 10^{-8}}{2 \times 10^4} =$

c) $\dfrac{2 \times 10^4}{4 \times 10^8} =$

Solutions to Practice Problem Set #3

Solution to Practice Problem 1: *Convert the following numbers to scientific notation:*

a)   53,000

> Start at the leftmost number, which is 5. From there, we have four digits. So, we rewrite 53,000 as 5.3×10^4.

b)   572,000,000,000

> Start at the leftmost number, which is 5. From there, we have 11 digits. So, we rewrite 572,000,000,000 as 5.72×10^{11}.

c) 0.00068

We count the number of zeros and get 3. Thus the power of 10 is −4 and we rewrite 0.00068 as 6.8×10^{-4}.

d) 0.0000000914

We count the number of zeros and get 7. Thus the power of 10 is −8 and we can rewrite 0.0000000914 as 9.14×10^{-8}.

e) 0.00003142

We count the number of zeros and get 4. Thus the power of 10 is −5 and we can rewrite 0.00003142 as 3.142×10^{-5}.

Solution to Practice Problem 2: *Convert the following numbers to ordinary decimal notation:*

a) 6.4×10^3

Here, 10 is raised to the power 3, so we need to move the decimal point three places to the right. We get: $6.4 \times 10^3 = 6400$.

b) 3.7×10^5

Here, 10 is raised to the power 5, so we need to move the decimal point five places to the right. We get: $3.7 \times 10^5 = 370,000$.

c) 2.735×10^7

Here, 10 is raised to the power 7, so we need to move the decimal point seven places to the right. We get: $2.735 \times 10^7 = 27,350,000$.

d) 5.4×10^{-6}

Here, 10 is raised to the power −6 so we need to move the decimal point six places to the left. We get: $5.4 \times 10^{-6} = 0.0000054$.

e) 9.778×10^{-9}

Here, 10 is raised to the power −9, so we need to move the decimal point nine places to the left. We get: $9.778 \times 10^{-9} = 0.000000009778$.

Solution to Practice Problem 3: *Find the following sums:*

a) $3.2 \times 10^4 + 5.2 \times 10^4 =$

We add 3.2 and 5.2 and multiply the product by 10^4, giving us 8.4×10^4.

b) $6.6 \times 10^5 + 8.2 \times 10^5 =$

We add 6.6 and 8.2 and multiply the product by 10^5, giving us 14.8×10^5. We rewrite this as 1.48×10^6.

c) $3.5 \times 10^8 + 4.1 \times 10^7 =$

We first rewrite one of the numbers so that it has the same power of 10 as the other number. It is generally easier to do the arithmetic when we rewrite the number with the higher power. Let's rewrite 3.5×10^8 as 35×10^7. Now we add the two numbers to get $35 \times 10^7 + 4.1 \times 10^7 = 39.1 \times 10^7$, which we then rewrite as 3.91×10^8.

Solution to Practice Problem 4: *Find the following differences:*

a) $5.2 \times 10^6 - 3.2 \times 10^6 =$

We subtract 3.2 from 5.2 and multiply the product by 10^6, giving us 2×10^6.

b) $4.9 \times 10^5 - 8.2 \times 10^5 =$

We subtract 8.2 from 4.9 and multiply the product by 10^5, giving us -3.3×10^5.

c) $5.1 \times 10^8 - 3.9 \times 10^7 =$

We first rewrite one of the numbers so that it has the same power of 10 as the other number. It is generally easier to do the arithmetic when we rewrite the number with the higher power. Let's rewrite 5.1×10^8 as 51×10^7. Now we can subtract the two numbers to get $51 \times 10^7 - 3.9 \times 10^7 = 47.1 \times 10^7$, which we rewrite as 4.71×10^8.

Solution to Practice Problem 5: *Find the following products:*

a) $\left(2.5 \times 10^3\right)\left(4 \times 10^5\right) =$

We multiply 2.5 and 4, and add the powers of 10. Here, we get $2.5 \cdot 4 = 10$, so $\left(2.5 \times 10^3\right)\left(4 \times 10^5\right) = 10 \times 10^8 = 1 \times 10^9$ (or just plain 10^9).

b) $\left(3.6 \times 10^6\right)\left(8 \times 10^7\right) =$

We multiply 3.6 and 8, and add the powers of 10. Here, we get $3.6 \cdot 8 = 28.8$, so $\left(3.6 \times 10^6\right)\left(8 \times 10^7\right) = 28.8 \times 10^{13} = 2.88 \times 10^{14}$.

c) $\left(9 \times 10^9\right)\left(1.6 \times 10^{-19}\right) =$

We multiply 9 and 1.6, and add the powers of 10. Here, we get $9 \cdot 1.6 = 14.4$, so $\left(9 \times 10^9\right)\left(1.6 \times 10^{-19}\right) = 14.4 \times 10^{-10} = 1.44 \times 10^{-9}$.

Solution to Practice Problem 6: *Find the following quotients:*

a) $\dfrac{2.5 \times 10^9}{5 \times 10^3} =$

Let's first rewrite this as $\dfrac{25 \times 10^8}{5 \times 10^3}$. Now we divide 25 by 5, and we subtract the

powers. We get $\dfrac{25}{5} = 5$, and $\dfrac{10^8}{10^3} = 10^{8-3} = 10^5$, which gives us $\dfrac{25 \times 10^8}{5 \times 10^3} = 5 \times 10^5$.

b) $\dfrac{4 \times 10^{-8}}{2 \times 10^4} =$

We divide 4 by 2, and we subtract the powers. We get $\dfrac{4}{2} = 2$, and

$\dfrac{10^{-8}}{10^4} = 10^{-8-4} = 10^{-12}$, which gives us $\dfrac{4 \times 10^{-8}}{2 \times 10^4} = 2 \times 10^{-12}$.

c) $\dfrac{2 \times 10^4}{4 \times 10^8} =$

We divide 2 by 4, and we subtract the powers. We get $\dfrac{2}{4} = 0.5$, and

$\dfrac{10^4}{10^8} = 10^{4-8} = 10^{-4}$, which gives us $\dfrac{2 \times 10^4}{4 \times 10^8} = 0.5 \times 10^{-4}$, which we rewrite as

5×10^{-5}.

UNIT FOUR

Graphs of Exponential Functions

Now that we are comfortable with exponential expressions, let's look at exponential functions. What is an exponential function? It is an equation that contains a variable in the exponent. These show up in many "real world" situations, such as compound interest, radioactive decay, population growth, and more. We will explore these topics later in the book. Here, we will look at the graphs of exponential functions, which is an excellent way to visualize exponential behavior.

First, let's think about linear growth. Suppose, for example, that we have a cookie shop and that we get $2 every time we sell a box of cookies. When we sell one box, we have $2. When we sell two boxes, we have $2 + 2 = 4. When we sell three boxes, we have $2 + 2 + 2 = 6. And so on. Our data looks like this:

x (boxes)	y (dollars)
0	0
1	2
2	4
3	6
4	8

Notice that every time x changes by 1, y changes by 2. This should make sense because the amount of money we get when we sell a box is the same every time. When x grows from 0 to 4, y grows from 0 to 8. We could represent the relationship between x and y by an equation of the form $y = mx$, where m is the *slope*. The slope describes the relationship between x and y, where every time x changes by a certain amount, y changes by a fixed amount. In this case, the equation is $y = 2x$ and the slope of the equation is 2.

If we graph the equation $y = 2x$, it looks like this:

Figure 1

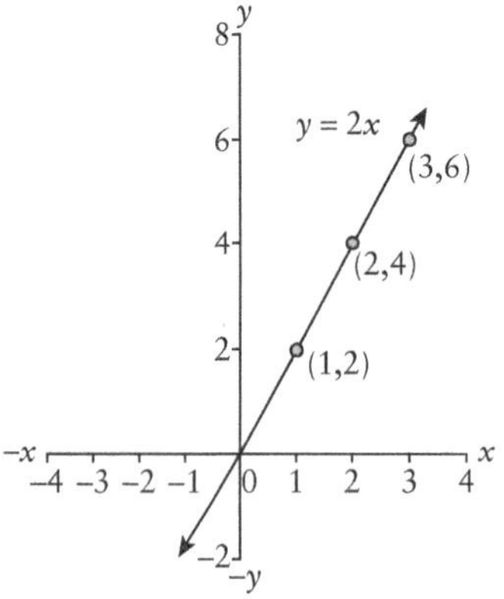

Now let's look at an exponential function. Suppose we have a life form that grows by splitting itself in half into two identical life forms, once an hour. Then those life forms split in half, and so on. The equation for this type of growth is $y = 2^x$. How did we get this? Note that after 1 hour, we have 2 life forms. After 2 hours, we have $2 \cdot 2 = 4$ life forms. After 3 hours, we have $2 \cdot 2 \cdot 2 = 8$ life forms. And so on. Our data looks like this:

x (hours)	y (life forms)
0	1
1	2
2	4
3	8
4	16

Notice the difference between this type of growth and linear growth. Here, when x changes by 1, y does not change by the same amount. In fact, when x increases by 1, y is multiplied by 2. Here, when x grows from 0 to 4, y grows from 0 to 16. It is growing much faster exponentially than it is linearly. For example, in the linear growth example, when x is 10, y is 20, whereas in the exponential growth example, when x is 10, y is 1024.

The graph of $y = 2^x$ looks like this:

Figure 2

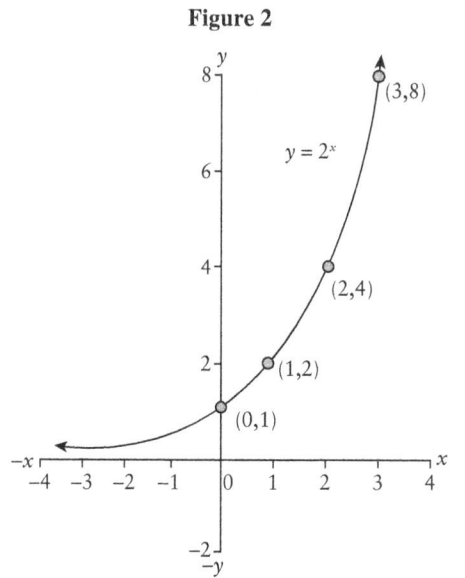

If we put the two graphs on the same axes, we can see how much faster the exponential graph increases relative to the linear one.

Figure 3

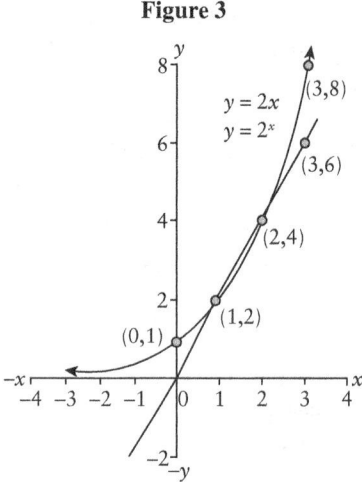

This is because, as each x value increases, the value of y in a linear graph increases at the same rate whereas the value of y in an exponential graph increases by ever greater amounts. For example, in the linear graph, when x is 5, y is 10, when x is 6, y is 12, and when x is 7, y is 14. For each increase of 1 in the value of x, y increased by 2. On the other hand, in the exponential graph, when x is 5, y is 32, when x is 6, y is 64, and when x is 7, y is 128. So, when x increased from 5 to 6, y increased by 32; when x increased from 6 to 7, y increased by 64.

This brings us to the essential features of exponential functions – they increase very rapidly (if x is greater than 1) because terms are being multiplied each time.

Let's look at another exponential function, to see the similarities between the functions.

Example 1: This time, we will graph $y = 3^x$. First, let's make a table of some values:

x	y
0	1
1	3
2	9
3	27
4	81

And here is what the graph looks like:

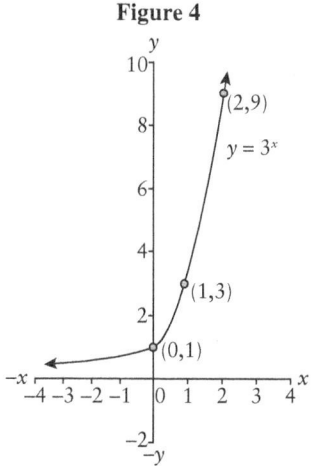

Figure 4

Notice the similarity in shape to the graph of $y = 2^x$.

So far, we have just examined values of y when x has nonnegative integer values. What happens when x is a negative number? Let's look at some values of $y = 3^x$:

x	y
-1	$\dfrac{1}{3}$
-2	$\dfrac{1}{9}$
-3	$\dfrac{1}{27}$
-4	$\dfrac{1}{81}$

Remember that $b^{-x} = \dfrac{1}{b^x}$, so as x gets more negative, the value of y gets closer to 0, but is still positive.

There are a couple of important aspects of an exponential graph to learn. First, all graphs of the form $y = b^x$, where $b > 1$, have the same basic shape. Second, notice that when $x = 0$, $y = 1$ because b raised to the power of 0 is always 1 (see Unit One). Third, the graphs have a horizontal asymptote of $y = 0$ (the x-axis). Finally, we can make x any value that we want, but y will always be a positive number. That is, the domain of an exponential function is $-\infty < x < \infty$, and the range is $y > 0$.

Example 2: Now let's look at the graph of $y = 2^x + 1$.

Figure 5

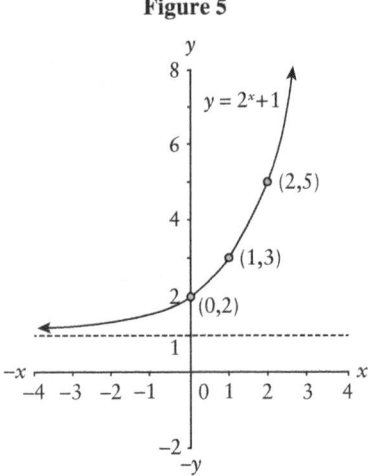

Note that the graph is identical to the graph of $y = 2^x$ but is shifted up one unit vertically. Thus, the horizontal asymptote is now $y = 1$, and the y-intercept is now $(0,2)$. In fact, for a graph of the form $y = b^x + d$, the horizontal asymptote is the line $y = d$ and the y-intercept is $(0, d + 1)$. For example, $y = 2^x + 4$ has a horizontal asymptote of $y = 4$ and a y-intercept at $(0,5)$.

Example 3: Let's look at the graph of $y = 3 \cdot 2^x$.

Figure 6

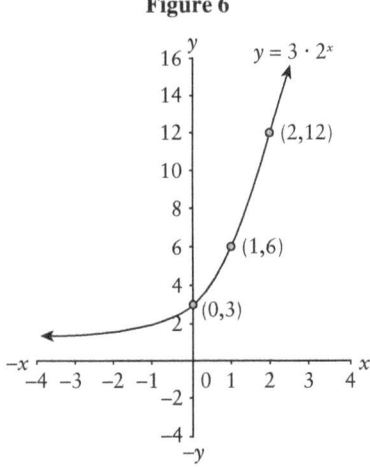

Here, the graph looks similar to $y = 2^x$ but it rises faster because each value of y is multiplied by 3. The y intercept is now $(0,3)$.

Example 4: Finally, let's look at the graph of $y = 2^{x-1}$.

Figure 7

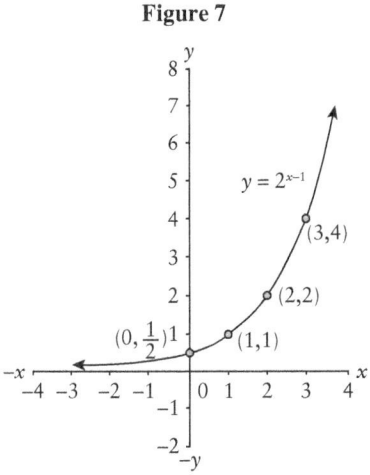

Here the graph again looks similar to $y = 2^x$ but has been shifted 1 unit to the right. Note that the graph goes through (1,1). This is because at $x = 1$, $y = 2^0 = 1$. So, instead of noting the *y-intercept*, we note the point that it has been shifted to.

Let's make a rule for graphing exponential functions.

If we have a function of the form $y = a \cdot b^{x-c} + d$ where $b > 1$, then d is the horizontal asymptote, and the y intercept is at $(0, a \cdot b^{-c} + d)$. The graph goes through the point $(c, a + d)$.

Figure 8

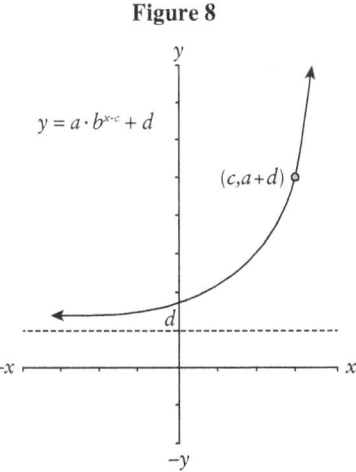

Now, let's look at a graph where $0 < b < 1$.

Example 5: Graph $y = \left(\dfrac{1}{2}\right)^x$. First, let's make a table of values:

x	y
−3	8
−2	4
−1	2
0	1
1	$\dfrac{1}{2}$
2	$\dfrac{1}{4}$

Notice that the same y values occur as we had for the graph $y = 2^x$, but now they are for negative x-values instead of positive ones. This will have the effect of reflecting the graph of $y = 2^x$ across the y axis.

Figure 9

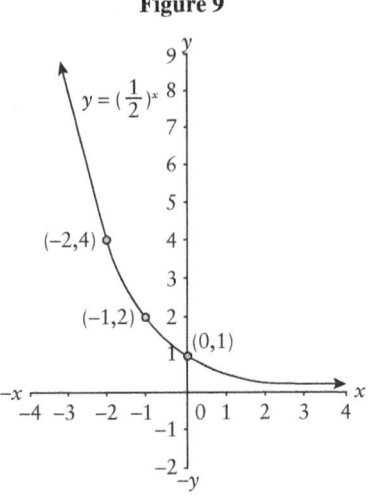

Note that the graph of $y = \left(\dfrac{1}{2}\right)^x$ has the same y intercept as the graph of $y = 2^x$, the same horizontal asymptote, and the same shape. Let's graph them together so we can compare them:

Figure 10

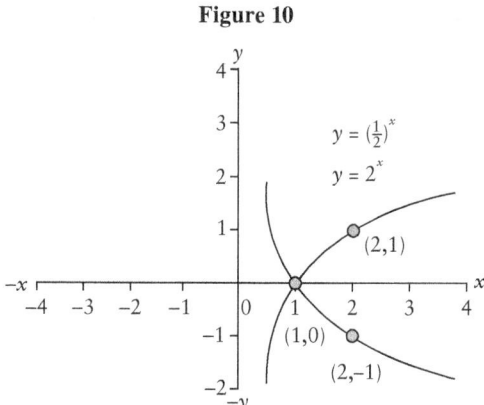

Now we have the basics of exponential graphs. Are you ready for some practice problems?

Practice Problem Set #4

Find the horizontal asymptote, the y intercept, and one other point. Graph each of the following functions:

1) $y = 4^x$

2) $y = 10^x$

3) $y = 4^x + 2$

4) $y = 10^x - 3$

5) $y = 2 \cdot 4^x + 1$

6) $y = \frac{1}{2} \cdot 10^{x-3}$

7) $y = 3 \cdot 4^{x+1} + 2$

8) $y = \left(\frac{1}{3}\right)^x$

9) $y = \left(\frac{1}{3}\right)^x + 2$

10) $y = \left(\frac{1}{3}\right)^{x-2} - 1$

Solutions to Practice Problem Set #4

Find the horizontal asymptote, the y intercept, and one other point. Graph each of the following functions:

1) $y = 4^x$

   The y intercept is $(0,1)$, it goes through $(1,4)$, and the horizontal asymptote is $y = 0$. The graph looks like this:

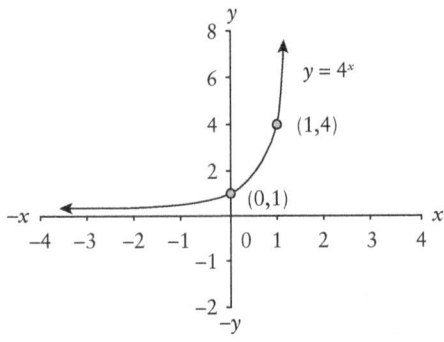

2) $y = 10^x$

   The y intercept is $(0,1)$, it goes through $(1, 10)$, and the horizontal asymptote is $y = 0$. The graph looks like this:

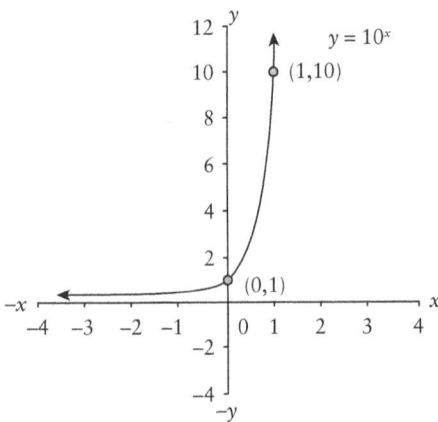

3) $y = 4^x + 2$

The y intercept is (0,3), it goes through (1,6), and the horizontal asymptote is $y = 2$. The graph looks like this:

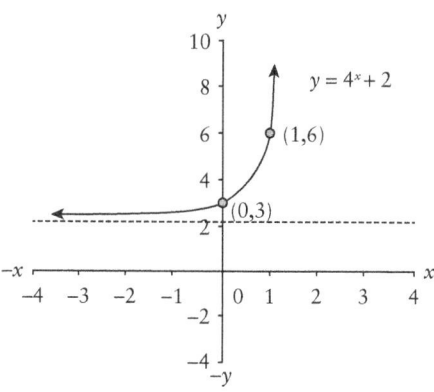

4) $y = 10^x - 3$

The y intercept is (0,–2), it goes through (1,7), and the horizontal asymptote is $y = -3$. The graph looks like this:

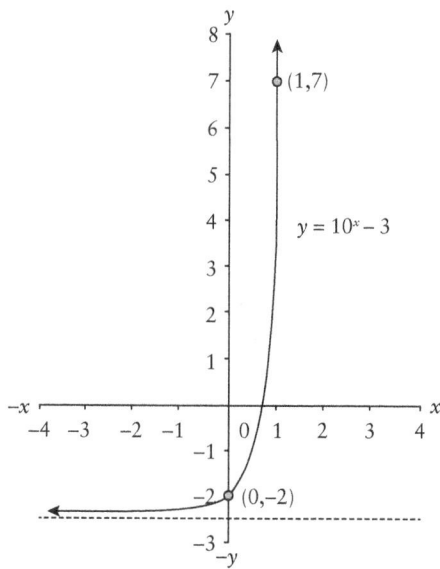

5) $y = 2 \cdot 4^x + 1$

The y intercept is $(0,3)$, it goes through $(1,9)$, and the horizontal asymptote is $y = 1$. The graph looks like this:

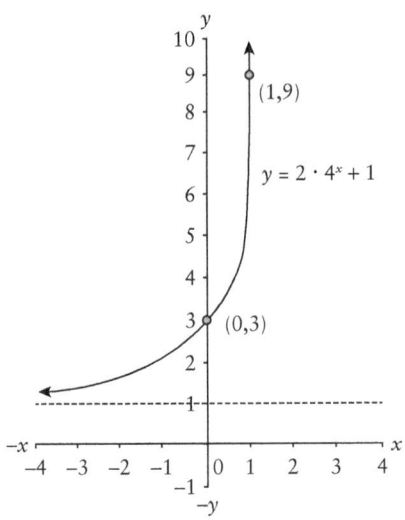

6) $y = \dfrac{1}{2} \cdot 10^{x-3}$

The y intercept is $\left(0, \dfrac{1}{2000}\right)$, it goes through $\left(3, \dfrac{1}{2}\right)$, and the horizontal asymptote is $y = 0$. The graph looks like this:

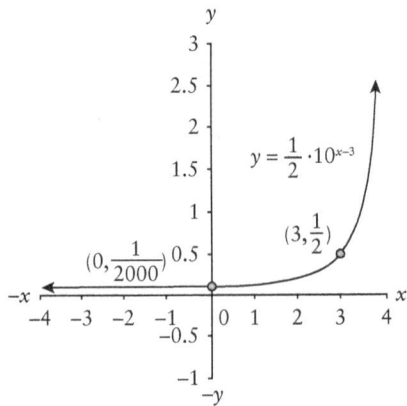

7) $y = 3 \cdot 4^{x+1} + 2$

The y intercept is $(0,14)$, it goes through $(-1,5)$, and the horizontal asymptote is $y = 2$. The graph looks like this:

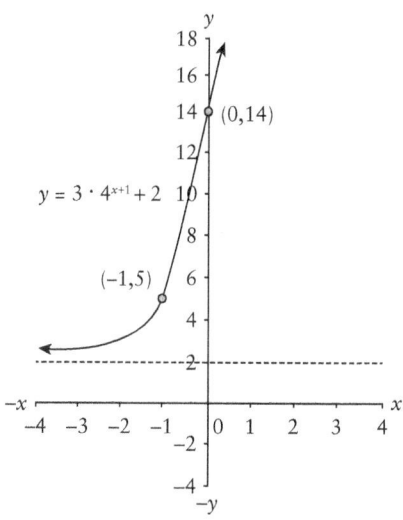

8) $y = \left(\dfrac{1}{3}\right)^x$

The y intercept is $(0,1)$, it goes through $(-1,3)$, and the horizontal asymptote is $y = 0$. The graph looks like this:

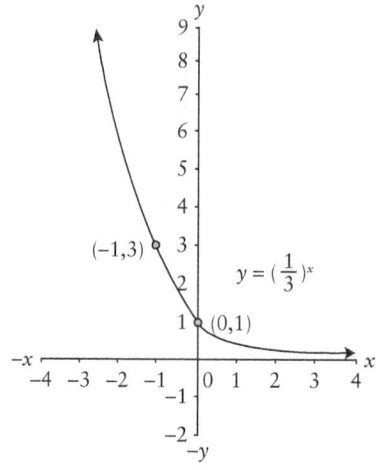

9) $y = \left(\dfrac{1}{3}\right)^{x} + 2$

The y intercept is $(0,3)$, it goes through $(-1,5)$, and the horizontal asymptote is $y = 2$. The graph looks like this:

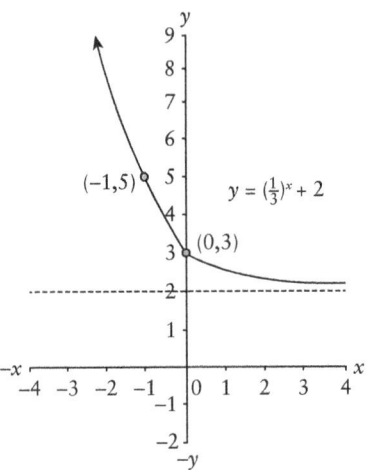

10) $y = \left(\dfrac{1}{3}\right)^{x-2} - 1$

The y intercept is $(0,8)$, it goes through $(2,0)$, and the horizontal asymptote is $y = -1$. The graph looks like this:

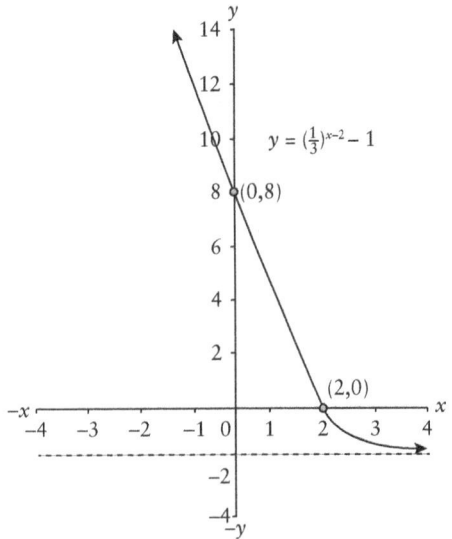

UNIT FIVE

Logarithms

Now that we have learned about exponentials, we are going to learn about logarithms. As we will see shortly, the two of them are closely related.

We know that $10^1 = 10$ and that $10^2 = 100$, so there must be a power of 10 that gives us 50. A good guess would be halfway between 1 and 2, namely $\frac{3}{2}$. Let's see what happens if we evaluate $10^{\frac{3}{2}}$. We know from the rules of exponents that

$$10^{\frac{3}{2}} = \sqrt{10^3} = \sqrt{1000} \approx 31.6.$$ So $\frac{3}{2}$ was a good guess but it is not correct. The actual answer is approximately 1.69. Don't worry about where we get the answer; it is beyond the scope of this book. The important point is that the number is indeed between 1 and 2, as we expected. We call this exponent the *logarithm of 50*, or *log 50* for short. In other words, *log 50* is the power that we raise 10 to in order to get 50. So *log 60* is the power that we raise 10 to in order to get 60 (≈ 1.78). Also, *log 73* ≈ 1.86 is the power that we raise 10 to in order to get 73, etc.

We write these as follows:

$\log 50 \approx 1.69$, which means that $10^{1.69} \approx 50$.
$\log 60 \approx 1.78$, which means that $10^{1.78} \approx 60$.
$\log 73 \approx 1.86$, which means that $10^{1.86} \approx 73$.

There is a difficulty, however. We know that $2^5 = 32$ and that $2^6 = 64$, so there must be a power of 2 that gives us 50. Furthermore, this power must be between 5 and 6 (it is approximately 5.64). We could also call this *log 50*. They can't both stand for the same thing so how do we distinguish between the *log 50* that is the power we use with a base of 10 and the *log 50* that is the power we use with a base of 2? Simple. We use a subscript with the *log* to indicate the base that we are using. We write $\log_{10} 50 \approx 1.69$ when we mean the power of 10 that gives us 50. We write $\log_2 50 \approx 5.64$ when we mean the power of 2 that gives us 50. So, if we write $\log_5 50 \approx 2.43$, it means that $5^{2.43} \approx 50$. This gives us a rule:

$$\boxed{\log_b x = a \text{ means that } b^a = x.}$$

Let's do some examples.

Example 1: What is $\log_7 49$?

We can write this as $\log_7 49 = x$, which means that we need to find the value of x where $7^x = 49$. The answer is $x = 2$.

Example 2: What is $\log_4 64$?

We can write this as $\log_4 64 = x$, which means that we need to find the value of x where $4^x = 64$. The answer is $x = 3$.

Example 3: What is $\log_{10} 1000$?

We can write this as $\log_{10} 1000 = x$, which means that we need to find the value of x where $10^x = 1000$. The answer is $x = 3$.

Example 4: Solve for x: $\log_5 x = 3$?

We can use our rule to write this as $5^3 = x$. The answer is $x = 125$.

We can see from these examples that sometimes a logarithm is a positive integer. Can we get other types of numbers for answers? Let's see.

Example 5: Solve for x: $\log_6 1 = x$.

Here, we can use our rule to get: $6^x = 1$. We know from our exponent rules that x must be 0. In fact, because any number (except 0) raised to the 0 power is 1, so we can make a rule:

$$\boxed{\log_b 1 = 0, \text{ where } b \text{ is any permissible base.}}$$

Now we know that a logarithm can be 0. Can we get a negative integer? Sure!

Example 6: Solve for x: $\log_2 \dfrac{1}{8} = x$.

Here we can use our rule to get: $2^x = \dfrac{1}{8}$. We know from our exponent rules that x must be -3.

Now we know that a logarithm can be a negative integer. Anything else?

Example 7: Solve for x: $\log_4 2 = x$.

Here we can use our rule to get: $4^x = 2$. We know from our exponent rules that x must be $\dfrac{1}{2}$.

Therefore, a logarithm can be a positive or negative integer, zero, or a rational number. Furthermore, a logarithm can be an irrational number. It just depends on what we are taking the log of. In other words, a logarithm can be any kind of real number. However, we *cannot* find the logarithm of any kind of number. Let's do an example.

Example 8: Solve for x: $\log_4 (-2) = x$.

This means that $4^x = -2$. But, there is no number that we can raise 4 to in order to get a negative number. The powers of 4 will always come out positive. Therefore, we cannot take the log of a negative number.

We also can't take the log of 0 (Try it and see!).

Remember:

If we are evaluating $y = \log_b x$, x must be a positive number; that is, the *domain* of the logarithm function is all positive real numbers. But, the logarithm of a number can be any real number; that is, the *range* of the logarithm function is all real numbers.

Logarithms that use base 10 are called *common logarithms* and when we write a *log* with no indicated base, it is presumed that we mean *log base 10*. That is, $\log x$ means $\log_{10} x$. Most calculators come equipped with a *log* button. When we use that button, we are finding the *common log* of a number.

Here are some of the basic properties of common logs:

$\log 1 = 0$ because $10^0 = 1$.

$\log 10 = 1$ because $10^1 = 10$.

$\log 10^x = x$ because x is the power that we raise 10 to in order to get 10^x.

$10^{\log x} = x$ because the log means that when we raise 10 to *log x*, we get x.

We could extend these properties to the logs of any base:

$\log_b 1 = 0$ because $b^0 = 1$.

$\log_b b = 1$ because $b^1 = b$.

$\log_b b^x = x$ because x is the power that we raise b to in order to get b^x.

$b^{\log_b x} = x$ because the log means that when we raise *10* to $\log_b x$, we get x.

Some calculators come equipped with a way to find logs of bases other than 10, but many do not. In the next unit, we will learn how to convert a log from one base to another, so that we can use the calculator to find logs of any base.

Are you ready for some practice problems?

Practice Problem Set #5

Evaluate the following logarithmic expressions without using a calculator:

1) $\log_8 8$

2) $\log_9 1$

3) $\log_2 32$

4) $\log_6 36$

5) $\log_{25} 5$

6) $\log_5 \sqrt[4]{5}$

7) $\log_{10} 10,000$

8) $\log_{10} 0.0001$

9) $\log_b b^3$

10) $\log_x \sqrt{x}$

Rewrite the following expressions in logarithmic form:

11) $6^2 = 36$

12) $4^3 = 64$

13) $10^{-2} = 0.01$

14) $2^{-3} = \dfrac{1}{8}$

15) $\sqrt[3]{125} = 5$

Rewrite the following expressions in exponential form:

16) $\log_9 81 = 2$

17) $\log_{10} 1,000,000 = 6$

18) $\log_3 243 = 5$

19) $\log_9 3 = \dfrac{1}{2}$

20) $\log_4 \dfrac{1}{64} = -3$

Solve the following expressions for x:

21) $\log_4 x = 4$

22) $\log 100,000 = x$

23) $\log x = -2$

24) $\log_2 x = \dfrac{1}{2}$

25) $\log_6 216 = x$

Solutions to Practice Problem Set #5

Evaluate the following logarithmic expressions without using a calculator:

1) $\log_8 8$

Remember that $\log_8 8$ means *what power do we raise 8 to in order to get 8?* We raise 8 to the power 1, so $\log_8 8 = 1$.

2) $\log_9 1$

Remember that $\log_9 1$ means *what power do we raise 9 to in order to get 1?* We raise 9 to the power 0, so $\log_9 1 = 0$.

3) $\log_2 32$

Remember that $\log_2 32$ means *what power do we raise 2 to in order to get 32?* We raise 2 to the power 5, so $\log_2 32 = 5$.

4) $\log_6 36$

Remember that $\log_6 36$ means *what power do we raise 6 to in order to get 36?*
We raise 6 to the power 2, so $\log_6 36 = 2$.

5) $\log_{25} 5$

Remember that $\log_{25} 5$ means *what power do we raise 25 to in order to get 5?*
We know that 5 is $\sqrt{25}$ and the power $\frac{1}{2}$ means *square root*, so $\log_{25} 5 = \frac{1}{2}$.

6) $\log_5 \sqrt[4]{5}$

Remember that $\log_5 \sqrt[4]{5}$ means *what power do we raise 5 to in order to get $\sqrt[4]{5}$?*
We know that the power $\frac{1}{4}$ means *fourth root*, 2, so $\log_5 \sqrt[4]{5} = \frac{1}{4}$.

7) $\log_{10} 10,000$

Remember that $\log_{10} 10,000$ means *what power do we raise 10 to in order to
get 10,000?* We raise 10 to the power 4, so $\log_{10} 10,000 = 4$.

8) $\log_{10} 0.0001$

Remember that $\log_{10} 0.0001$ means *what power do we raise 10 to in order to
get 0.0001?* We raise 10 to the power -4, so $\log_{10} 0.0001 = -4$.

9) $\log_b b^3$

Remember that $\log_b b^3$ means *what power do we raise b to in order to get b^3?*
We raise b to the power 3, so $\log_b b^3 = 3$.

10) $\log_x \sqrt{x}$

Remember that $\log_x \sqrt{x}$ means *what power do we raise x to in order to get $\sqrt{x}$?*
We raise x to the power $\frac{1}{2}$, so $\log_x \sqrt{x} = \frac{1}{2}$.

Rewrite the following expressions in logarithmic form:

11) $6^2 = 36$

Remember our rule that $\log_b x = a$ means that $b^a = x$. Here we get: $\log_6 36 = 2$.

12) $4^3 = 64$

Remember our rule that $\log_b x = a$ means that $b^a = x$. Here we get: $\log_4 64 = 3$.

13) $10^{-2} = 0.01$

Remember our rule that $\log_b x = a$ means that $b^a = x$. Here we get: $\log_{10} 0.01 = -2$.

14) $2^{-3} = \frac{1}{8}$

Remember our rule that $\log_b x = a$ means that $b^a = x$. Here we get: $\log_2 \frac{1}{8} = -3$.

15) $\sqrt[3]{125} = 5$

First, let's rewrite $\sqrt[3]{125}$ as $125^{\frac{1}{3}}$. Now we can use our rule that $\log_b x = a$ means that $b^a = x$. Here we get: $\log_{125}(5) = \dfrac{1}{3}$.

Rewrite the following expressions in exponential form:

16) $\log_9 81 = 2$

Remember our rule that $\log_b x = a$ means that $b^a = x$. Here we get: $9^2 = 81$.

17) $\log_{10} 1,000,000 = 6$

Remember our rule that $\log_b x = a$ means that $b^a = x$. Here we get: $10^6 = 1,000,000$.

18) $\log_3 243 = 5$

Remember our rule that $\log_b x = a$ means that $b^a = x$. Here we get: $3^5 = 243$.

19) $\log_9 3 = \dfrac{1}{2}$

Remember our rule that $\log_b x = a$ means that $b^a = x$. Here we get: $9^{\frac{1}{2}} = 3$.

20) $\log_4 \dfrac{1}{64} = -3$

Remember our rule that $\log_b x = a$ means that $b^a = x$. Here we get: $4^{-3} = \dfrac{1}{64}$.

Solve the following expressions for x:

21) $\log_4 x = 4$

Remember our rule that $\log_b x = a$ means that $b^a = x$. Here we get: $4^4 = x$, so $x = 256$.

22) $\log 100,000 = x$

Remember our rule that $\log_b x = a$ means that $b^a = x$. Note that there is no base indicated for the log, so the base is 10. Here we get: $10^x = 100,000$, so $x = 5$.

23) $\log x = -2$

Remember our rule that $\log_b x = a$ means that $b^a = x$. Note that there is no base indicated for the log, so the base is 10. Here we get: $10^{-2} = x$, so $x = \dfrac{1}{10^2} = \dfrac{1}{100}$.

24) $\log_2 x = \dfrac{1}{2}$

Remember our rule that $\log_b x = a$ means that $b^a = x$. Here we get: $2^{\frac{1}{2}} = x$, so $x = \sqrt{2}$.

25) $\log_6 216 = x$

Remember our rule that $\log_b x = a$ means that $b^a = x$. Here we get: $6^x = 216$, so $x = 3$.

UNIT SIX
Log Laws

Now that we have learned what logarithms are, let us learn how to use them. We will look at a variety of problems involving logarithms, and there are four basic rules we call the *Log Laws*, that will enable us to manipulate the logarithms and to solve these problems.

Our first Log Law deals with products. When we have the log of two functions that are multiplied together, we can take the log of each function separately and add the results. In other words, if we are given two functions A and B, then $\log_b (AB) = \log_b A + \log_b B$. Notice that this is similar to the product rule for exponents, namely that $x^A x^B = x^{A+B}$. This is not a coincidence!

Here is where it comes from:

Let $x = \log_b A$ and $y = \log_b B$. We can now use the definition of a logarithm to rewrite these as $b^x = A$ and $b^y = B$. Then, $AB = b^x \cdot b^y = b^{x+y}$. Now we can again use the definition of a logarithm to rewrite this as $\log_b (AB) = x + y$. Therefore, because $x = \log_b A$ and $y = \log_b B$, we get $\log_b (AB) = \log_b A + \log_b B$.

Example 1: Expand $\log(5xy)$.

Using the log law that we just learned, we can rewrite $\log(5xy)$ as $\log 5 + \log x + \log y$.

Our second Log Law deals with quotients in log equations. When we have the log of two functions that are divided, we can take the log separately and subtract the results. In other words, if we are given two functions A and B, then $\log_b \left(\dfrac{A}{B} \right) = \log_b A - \log_b B$. Notice that this is similar to the quotient rule for exponents, namely that $\dfrac{x^A}{x^B} = x^{A-B}$. Try to derive it yourself, following the method that we did for the product law above.

Example 2: Expand $\log \dfrac{3}{y}$.

Using the log law that we just learned, we can rewrite $\log \dfrac{3}{y}$ as $\log 3 - \log y$.

The third Log Law deals with powers. When we have the log of a function that is raised to a power, we can multiply the power by the log of the base. In other words, $\log_b A^B = B \log_b A$.

Again, we will leave it to you to derive. Use the same method as for the Product Law.

Here are the three Log Laws again:

Product Law:	$\log_b (AB) = \log_b A + \log_b B$
Quotient Law:	$\log_b \left(\dfrac{A}{B} \right) = \log_b A - \log_b B$
Power Law:	$\log_b (A^B) = B \log_b A$

Let's do some examples.

Example 3: Use the Log Laws to expand $\log \left(\dfrac{A^2 B}{C^3} \right)$.

First, let's use the Quotient Law to rewrite this as $\log \left(\dfrac{A^2 B}{C^3} \right) = \log A^2 B - \log C^3$.

Next, use the Product Law to rewrite the first term: $\log(A^2 B) - \log(C^3) = \log(A^2) + \log B - \log(C^3)$. Finally, let's use the Power Law to rewrite the first and third terms: $\log A^2 + \log B - \log C^3 = 2\log A + \log B - 3\log C$.

Example 4: Use the Log Laws to expand $\log(B\sqrt{A})$.

First, let's rewrite the log as the $\log \left(BA^{\frac{1}{2}} \right)$. Now we can use the Product Law to rewrite this as $\log \left(BA^{\frac{1}{2}} \right) = \log B + \log \left(A^{\frac{1}{2}} \right)$. Finally, let's use the Power Law: $\log \left(BA^{\frac{1}{2}} \right) = \log B + \dfrac{1}{2}\log A$.

Example 5: Use the Log Laws to write the expression as a single logarithm: $5\log A + 3\log B - \dfrac{1}{2}\log C$.

First, let's use the Power Law: $5\log A + 3\log B - \dfrac{1}{2}\log C = \log(A^5) + \log(B^3) - \log\left(C^{\frac{1}{2}} \right)$. Next, let's use the Product Law to combine the first two terms: $\log(A^5) + \log(B^3) - \log\left(C^{\frac{1}{2}} \right) = \log(A^5 B^3) - \log\left(C^{\frac{1}{2}} \right)$. Now we can use the Quotient Law to rewrite this as $\log(A^5 B^3) - \log\left(C^{\frac{1}{2}} \right) = \log\left(\dfrac{A^5 B^3}{C^{\frac{1}{2}}} \right)$, which can also be written as $\log\left(\dfrac{A^5 B^3}{\sqrt{C}} \right)$.

Here is another type of problem that uses the Log Laws.

Example 6: If $\log x = 3$, $\log y = 4$ and $\log z = 6$, find $\log\left(\dfrac{x^3\sqrt{y}}{z^2}\right)$.

First, let's use the Log Laws to expand the log. We can use the Quotient Law to get: $\log\left(\dfrac{x^3\sqrt{y}}{z^2}\right) = \log\left(x^3\sqrt{y}\right) - \log\left(z^2\right)$. Then, we can use the Product Law to get: $\log\left(x^3\sqrt{y}\right) - \log\left(z^2\right) = \log\left(x^3\right) + \log\sqrt{y} - \log\left(z^2\right)$. Next, we can use the Power Law to get: $\log\left(x^3\right) + \log\sqrt{y} - \log\left(z^2\right) = 3\log x + \dfrac{1}{2}\log y - 2\log z$. Finally, now that we have expanded the logarithm, we can plug in the values of the individual logs to get: $3\log x + \dfrac{1}{2}\log y - 2\log z = 3(3) + \dfrac{1}{2}(4) - 2(6) = -1$.

There is a fourth rule that we will learn that is not one of the *Laws*, but is usually learned at the same time. It is called the *Change of Base Rule*. It enables us to take a log with a given base and write it in terms of a different base. This can be very handy when finding logs on the calculator when the base is not 10, or when combining logs with more than one base.

$$\text{Change of Base Rule: } \log_b x = \frac{\log_a x}{\log_a b}$$

Example 7: Use the Change of Base Rule to evaluate $\log_8 32$.

We can use the Change of Base Rule and convert this into a log with a base 2: $\log_8 32 = \dfrac{\log_2 32}{\log_2 8}$. We know that $\log_2 32 = 5$ because $2^5 = 32$, and we know that $\log_2 8 = 3$ because $2^3 = 8$, so $\dfrac{\log_2 32}{\log_2 8} = \dfrac{5}{3}$.

We also could have done the above example by converting the log into a log with a base 10: $\log_8 32 = \dfrac{\log 32}{\log 8}$. In fact, we could convert the log into any base that we want, as long as we use the same base for the numerator and the denominator. If we plug that into a calculator, we get $\dfrac{5}{3}$. Although we didn't need to use a calculator here we will only be able to get a numerical answer using a calculator in many problems.

Note that when we find logs with a calculator, we get a decimal approximation of the answer, not an *exact* one. For example, if we find $\log 2$ with a calculator, we get $\log 2 \approx 0.3010$. Most of the time, this will be close enough, but it is not exact. The *exact* value is $\log 2$. This is true for all irrational numbers, not just logarithms.

Time to practice!

Practice Problem Set #6

Use the Log Laws to expand the following logarithms:

1) $\log\dfrac{8}{y}$

2) $\log(7xyz)$

3) $\log\left(\dfrac{5A}{B}\right)$

4) $\log(x^4 y^6)$

5) $\log\left(\dfrac{y\sqrt[3]{x}}{w^5}\right)$

6) $\log\sqrt{\dfrac{x^4}{y^7}}$

Use the Log Laws to write the following as a single logarithm:

7) $\log 5 - \log 11$

8) $4\log A + 3\log B - 10\log C$

9) $9\log w - 4\log z + 2\log x$

10) $\dfrac{1}{4}(\log x + 3\log y - 2\log z)$

11) $8\log(xy) - 5\log(yz)$

12) $3\log(x^2 y) + 4\log(yz^3)$

Use the Change of Base Rule and your calculator to evaluate the following logarithms:

13) $\log_6 32$

14) $\log_7 9$

15) $\log_2 50$

Solutions to Practice Problem Set #6

Use the Log Laws to expand the following logarithms:

1) $\log\dfrac{8}{y}$

Here, we can use the Quotient Law, which says that $\log\dfrac{A}{B} = \log A - \log B$.

We get: $\log\dfrac{8}{y} = \log 8 - \log y$

2) $\log(7xyz)$

Here, we can use the Product Law, which says that $\log(AB) = \log A + \log B$. Of course, if we have more than two terms multiplied together, we can add all of their logs.

We get: $\log(7xyz) = \log 7 + \log x + \log y + \log z$

3) $\log\left(\dfrac{5A}{B}\right)$

First, we can use the Quotient Law: $\log\left(\dfrac{5A}{B}\right) = \log(5A) - \log B$. Next, we use the Product Law to get: $\log(5A) - \log B = \log 5 + \log A - \log B$.

4) $\log\left(x^4 y^6\right)$

First, we can use the Product Law: $\log\left(x^4 y^6\right) = \log\left(x^4\right) + \log\left(y^6\right)$. Next, we can use the Power Law, which says that $\log A^B = B \log A$. We get: $\log\left(x^4\right) + \log\left(y^6\right) = 4\log x + 6\log y$.

5) $\log\left(\dfrac{y\sqrt[3]{x}}{w^5}\right)$

First, we can use the Quotient Law: $\log\dfrac{y\sqrt[3]{x}}{w^5} = \log y\sqrt[3]{x} - \log w^5$. Next, we can use the Product Law: $\log y\sqrt[3]{x} - \log w^5 = \log y + \log \sqrt[3]{x} - \log w^5$. Finally, we can use the Power Law: $\log y + \log \sqrt[3]{x} - \log w^5 = \log y + \dfrac{1}{3}\log x - 5\log w$.

6) $\log\sqrt{\dfrac{x^4}{y^7}}$

First, let's rewrite the radical as a power: $\log\sqrt{\dfrac{x^4}{y^7}} = \log\left(\dfrac{x^4}{y^7}\right)^{\frac{1}{2}}$. Now, let's use the Power Law: $\log\left(\dfrac{x^4}{y^7}\right)^{\frac{1}{2}} = \dfrac{1}{2}\log\left(\dfrac{x^4}{y^7}\right)$. Next, let's use the Quotient Law: $\dfrac{1}{2}\log\left(\dfrac{x^4}{y^7}\right) = \dfrac{1}{2}\left(\log x^4 - \log y^7\right)$. Finally, let's use the Power Law again to get: $\dfrac{1}{2}\left(\log x^4 - \log y^7\right) = \dfrac{1}{2}\left(4\log x - 7\log y\right)$.

Use the Log Laws to write the following as a single logarithm:

7) $\log 5 - \log 11$

Here, we can use the Quotient Law, which says that $\log\left(\dfrac{A}{B}\right) = \log A - \log B$.

We get: $\log 5 - \log 11 = \log\left(\dfrac{5}{11}\right)$.

8) $4\log A + 3\log B - 10\log C$

First, we can use the Power Law, which says that $\log\left(A^B\right) = B\log A$. We get: $4\log A + 3\log B - 10\log C = \log\left(A^4\right) + \log\left(B^3\right) - \log\left(C^{10}\right)$. Next, we can use the Product Law, which says that $\log(AB) = \log A + \log B$, to get: $\log\left(A^4\right) + \log\left(B^3\right) - \log C^{10} = \log\left(A^4 B^3\right) - \log\left(C^{10}\right)$. Finally, we can use the Quotient Law: $\log\left(A^4 B^3\right) - \log\left(C^{10}\right) = \log\left(\dfrac{A^4 B^3}{C^{10}}\right)$.

9) $9\log w - 4\log z + 2\log x$

First, we can use the Power Law: $9\log w - 4\log z + 2\log x = \log\left(w^9\right) - \log\left(z^4\right) + \log\left(x^2\right)$. Next, we can use the Product Law to get: $\log\left(w^9\right) - \log\left(z^4\right) + \log\left(x^2\right) = \log\left(w^9 x^2\right) - \log\left(z^4\right)$. Notice that we combined the first and the third logs. We don't have to take them in order! Finally, we can use the Quotient Law: $\log\left(w^9 x^2\right) - \log\left(z^4\right) = \log\left(\dfrac{w^9 x^2}{z^4}\right)$.

10) $\dfrac{1}{4}\left(\log x + 3\log y - 2\log z\right)$

First, we can use the Power Law for the terms inside the parentheses. We will leave the term outside the parentheses for last. We get: $\dfrac{1}{4}\left(\log x + 3\log y - 2\log z\right) = \dfrac{1}{4}\left[\log x + \log\left(y^3\right) - \log\left(z^2\right)\right]$. Next, we can use the Product Law to get: $\dfrac{1}{4}\left[\log x + \log\left(y^3\right) - \log\left(z^2\right)\right] = \dfrac{1}{4}\left[\log\left(xy^3\right) - \log\left(z^2\right)\right]$. Now, we can use the Quotient Law to get: $\dfrac{1}{4}\left[\log\left(xy^3\right) - \log\left(z^2\right)\right] = \dfrac{1}{4}\left[\log\left(\dfrac{xy^3}{z^2}\right)\right]$. Finally, we can use the Power Law again to get: $\dfrac{1}{4}\left[\log\left(\dfrac{xy^3}{z^2}\right)\right] = \log\left[\left(\dfrac{xy^3}{z^2}\right)^{\frac{1}{4}}\right]$.

11) $8\log(xy) - 5\log(yz)$

First, we can use the Power Law on the two logs separately: $8\log(xy) - 5\log(yz) = \log\left[(xy)^8\right] - \log\left[(yz)^5\right]$. Next, we can use the Quotient Law to get: $\log(xy)^8 - \log(yz)^5 = \log\left[\dfrac{(xy)^8}{(yz)^5}\right]$. Although we could stop here, let's simplify the logarithm. Distribute the powers to their terms: $\log\left[\dfrac{(xy)^8}{(yz)^5}\right] = \log\left(\dfrac{x^8 y^8}{y^5 z^5}\right)$.

Then combine like terms: $\log\left(\dfrac{x^8 y^8}{y^5 z^5}\right) = \log\left(\dfrac{x^8 y^3}{z^5}\right)$.

12) $3\log\left(x^2 y\right) + 4\log\left(yz^3\right)$

First, we can use the Power Law on the two logs separately: $3\log\left(x^2 y\right) + 4\log\left(yz^3\right) = \log\left[\left(x^2 y\right)^3\right] + \log\left[\left(yz^3\right)^4\right]$. Next, we can use the Product Law to get: $\log\left(x^2 y\right)^3 + \log\left(yz^3\right)^4 = \log\left[\left(x^2 y\right)^3 \left(yz^3\right)^4\right]$. Although we could stop here, let's simplify the logarithm. Distribute the powers to their terms: $\log\left[\left(x^2 y\right)^3 \left(yz^3\right)^4\right] = \log\left[\left(x^6 y^3\right)\left(y^4 z^{12}\right)\right]$. Then combine like terms: $\log\left[\left(x^6 y^3\right)\left(y^4 z^{12}\right)\right] = \log\left(x^6 y^7 z^{12}\right)$.

Use the Change of Base Rule and your calculator to evaluate the following logarithms:

13) $\log_6 32$

The Change of Base Rule says that $\log_b x = \dfrac{\log_a x}{\log_a b}$. We can change the logarithms to any base we want, but if we change the base of the logarithms to base 10, we can use a calculator. We get: $\log_6 32 = \dfrac{\log 32}{\log 6} \approx \dfrac{1.5051}{0.7782} \approx 1.9341$.

14) $\log_7 9$

The Change of Base Rule says that $\log_b x = \dfrac{\log_a x}{\log_a b}$. We can change the logarithms to any base we want, but if we change the base of the logarithms to base 10, we can use a calculator. We get: $\log_7 9 = \dfrac{\log 9}{\log 7} \approx \dfrac{0.9542}{0.8451} \approx 1.1291$.

15) $\log_2 50$

The Change of Base Rule says that $\log_b x = \dfrac{\log_a x}{\log_a b}$. We can change the logarithms to any base we want, but if we change the base of the logarithms to base 10, we can use a calculator. We get: $\log_2 50 = \dfrac{\log 50}{\log 2} \approx \dfrac{1.6990}{0.3010} \approx 5.6445$.

By the way, in the previous unit we told you that $\log_2 50 \approx 5.6439$. Now you know where that came from!

UNIT SEVEN

Exponential Growth and Natural Logarithms

Suppose that you want to put some money in a bank account, where it will earn interest. Banks generally pay what is called *compound interest.* This means that we earn a certain rate of interest on our money over a specified time period, and, if we leave our money in the bank for longer than that time period, we get more interest, based not just on our original deposit but also on the interest that we have earned. We will see what this means in just a moment. First, let's make sure that we understand how to compute interest. Suppose that we put $1000 in a bank and that we earn 10% interest after one year. Then, at the end of the year, the bank pays us $0.10 \cdot \$1000 = \$100$, or $100 interest. Now let's see how compound interest works.

Example 1: We deposit $1000 in a bank that pays 10% interest annually. How much will we have after three years?

At the end of the first year, we will earn $0.10 \cdot \$1000 = \$100$ in interest. In the second year, we now have $\$1000 + \$100 = \$1100$ earning interest in the bank. So, at the end of the second year, we will earn $0.10 \cdot \$1100 = \$110$ interest. In the third year, we now $\$1100 + \$110 = \$1210$ in the bank earning interest. So, at the end of the third year, we will earn $0.10 \cdot \$1210 = \$121$ in interest. Therefore, at the end of three years, we will have $\$1210 + \$121 = \$1331$ in the bank. Let's make a table to help us visualize the calculations.

Year	Beginning of Year	Interest Earned	End of Year
1	$1000	$0.10 \cdot \$1000 = \$100$	$\$1000 + \$100 = \$1100$
2	$1100	$0.10 \cdot \$1100 = \$110$	$\$1100 + \$110 = \$1210$
3	$1210	$0.10 \cdot \$1210 = \$121$	$\$1210 + \$121 = \$1331$

Notice that if we had not left the interest in the bank each year, we would have found the total amount of money after three years by finding 10% of $1000 and multiplying it by 3, which would give us $1300 (this is called *simple interest*). So, by leaving the interest in the bank, we earned extra money. This may not seem like a lot but over time, it makes a big difference. This kind of growth is also called *exponential growth*, and it is a phenomenon that occurs frequently in many areas of natural science, as well as in finance.

Now that we have seen a simple example, let's find a rule for exponential growth. We will repeat the previous example but using variables in place of the numbers.

Example 2: We deposit \$1000.00 in a bank that pays R interest annually. How much will we have after three years?

At the end of the first year, we will earn $R \cdot P$ in interest. In the second year, we now have $P + R \cdot P$ in the bank earning interest. If we factor out the P, we get $P(1 + R)$. At the end of the second year, we will earn $R \cdot P(1 + R)$ in interest. In the third year, we now have $P(1 + R) + \left[R \cdot P(1 + R)\right]$ in the bank earning interest. If we factor out the $P(1 + R)$ term, we get $\left[P(1 + R)\right](1 + R)$, which we can simplify to $P(1 + R)^2$. At the end of the third year, we will earn $R \cdot P(1 + R)^2$ in interest. Therefore, at the end of three years, we will have $P(1 + R)^2 + R \cdot P(1 + R)^2$ in the bank. If we factor out the $P(1 + R)^2$ term, we get $\left[P(1 + R)^2\right](1 + R)$, which we can simplify to $P(1 + R)^3$. Let's make a table to help us visualize the calculations.

Year	Beginning of Year	Interest Earned	End of Year
1	1000	$R \cdot 1000$	$1000(1 + R)$
2	$1000(1 + R)$	$R \cdot 1000(1 + R)$	$1000(1 + R)^2$
3	$1000(1 + R)^2$	$R \cdot 1000(1 + R)^2$	$1000(1 + R)^3$

We can see a pattern forming. After one year, we have $1000(1 + R)$. After two years, we have $1000(1 + R)^2$. After three years, we have $1000(1 + R)^3$. So what will happen after T years? We will have $1000(1 + R)^T$. Of course, there is nothing special about the \$1000 that we are depositing. It could just as easily be P dollars. This now leads us to a formula for exponential growth:

> If we deposit P dollars in the bank, at a rate of R % annually, for T years, we end up with an amount F, where $F = P(1 + R)^T$.

Let's check our formula. In Example 1, P was \$1000, R was 10%, and T was 3 years. According to the formula, we should get $F = 1000(1 + 0.1)^3 = 1331$. So our formula works!

Notice that we used 0.1 in the formula for R, not 10. Remember to convert the percentage into a decimal. Otherwise, you will end up with a nonsensical answer. Let's do another example.

Example 3: We deposit \$1000 in the bank for ten years at 6% interest annually. How much will we have after ten years? Compare the answer to how much we would have had if we had earned simple interest instead of compound interest.

Let's use our formula. Here, P is 1000, R is 0.06, and T is 10. Plugging in, we get: $F = 1000(1 + 0.06)^{10} = 1790.85$, or \$1790.85. What would we have had with simple interest? We would have earned $0.06 \cdot 1000 = 60$ a year for ten years, giving us \$1600. That's a big difference!

Notice that we have been receiving the interest once a year. Suppose that instead of paying us 6% once a year, the bank offered to pay us half the interest twice a year, that is, 3% semi-annually. Let's see what happens.

Example 4: We deposit $1000 in the bank for ten years at 6% interest, compounded semi-annually. How much will we have after ten years?

Let's use our formula, with a small change. Here, P is 1000, R is 0.06, and T is 10. But, in order for the answer to come out correctly, we need to divide the rate by 2 and multiply the time by 2. Why? Because we are earning 3% each half-year period, and we are receiving it for 20 half-year periods. Our formula becomes $F = P\left(1 + \dfrac{R}{2}\right)^{2T}$. Plugging in, we get: $F = 1000$ $\left(1 + \dfrac{0.06}{2}\right)^{2 \cdot 10} = 1000(1.03)^{20} = 1806.11$, or $1806.11.

Notice that earning half of the interest twice a year gives us more money than earning the full interest once a year.

What would happen if we received the interest quarterly; that is, four times a year?

Example 5: We deposit $1000 in the bank for ten years at 6% interest, compounded quarterly. How much will we have after ten years?

Again, we need to make a small change to our formula. Here, P is 1000, R is 0.06, and T is 10. But, in order for the answer to come out correctly, we need to divide the rate by 4 and multiply the time by 4. Our formula becomes $F = P\left(1 + \dfrac{R}{4}\right)^{4T}$.

Plugging in, we get: $F = 1000\left(1 + \dfrac{0.06}{4}\right)^{4 \cdot 10} = 1000(1.015)^{40} = 1814.02$, or $1814.02.

Now we made more than if we compounded the interest semi-annually.

What do you think happens if the interest compounds every month?

Example 6: We deposit $1000 in the bank for ten years at 6% interest, compounded monthly. How much will we have after ten years?

Again, we need to make a small change to our formula. Here, P is 1000, R is 0.06, and T is 10. But, in order for the answer to come out correctly, we need to divide the rate by 12 and multiply the time by 12. Our formula becomes $F = P\left(1 + \dfrac{R}{12}\right)^{12T}$.

Plugging in, we get: $F = 1000\left(1 + \dfrac{0.06}{12}\right)^{12 \cdot 10} = 1000(1.005)^{120} = 1819.40$, or $1819.40.

Let's notice a couple of things. First, the more often the interest compounds per year, the more money we make. Second, we divided the interest rate by the number of times the interest compounds in a year and we multiplied the time by the same number. This gives us a more general version of the formula.

If we deposit P dollars in the bank, at a rate of R % annually, for T years, compounded N times a year, we end up with an amount F, where

$$F = P\left(1 + \frac{R}{N}\right)^{NT}.$$

Suppose we went to a bank that offered to pay us interest continuously? Then N would become infinite and our formula wouldn't work. This was a question that puzzled mathematicians for a long time and was finally answered by Leonhard Euler. He found that if we write the term $\left(1 + \frac{1}{N}\right)^N$ and we let N become infinity (this is called "finding the limit"), then we get a constant, which was named e, in his honor. We will find that e is a very important number and shows up in all sorts of mathematical contexts, just as π does. The value of e is approximately 2.71828. As with π, it is irrational, so we generally work with an approximation, or we write answers in terms of e. So, if the interest is compounded continuously, our formula changes:

If we deposit P dollars in the bank, at a rate of R % annually, for T years, compounded continuously, we end up with an amount F, where $F = Pe^{RT}$.

Let's do an example.

Example 7: We deposit \$1000 in the bank for ten years at 6% interest, compounded continuously. How much will we have after ten years?

Here, P is 1000, R is 0.06, and T is 10. Plugging in, we get: $F = 1000e^{0.06 \cdot 10} = 1822.12$, or \$1822.12.

As we expect, we get the most money when the interest is compounded most frequently.

We have been using money as a way to think about exponential growth but, of course, many different phenomena could display this type of growth. Let's do an example.

Example 8: There is a pond that has algae growing in it. Every day, there is 5% more algae than the day before. If on the first day, there is ten square inches of algae on the pond, how much will there be after one week?

Which formula should we use? Notice that the algae is growing once a day or, more realistically, we measure its growth once a day. Then we can use our basic growth formula. Here, P is 10, R is 0.05, and T is 7. Plugging in, we get: $F = 10(1 + 0.05)^7 = 14.07$ square inches.

As we said above, e is a very important number and will show up in many mathematical applications. In fact, it is so important and so prevalent, that there is a special type of logarithm that uses e as its base. It is called the *natural logarithm* and, instead of being written as $\log_e x$, it is written as $\ln x$. So, using the definition of a logarithm that we learned earlier, $\ln A = B$ means that $e^B = A$.

All of the rules that we learned about logarithms apply, of course, to natural logarithms, namely:

$\ln 1 = 0$

$\ln e = 1$

$\ln e^x = x$

$e^{\ln x} = x$

Product Law: $\ln(AB) = \ln A + \ln B$

Quotient Law: $\ln \dfrac{A}{B} = \ln A - \ln B$

Power Law: $\ln(A^B) = B \ln A$

Change of Base Rule: $\log_b x = \dfrac{\ln x}{\ln b}$

Notice that the last item is the Change of Base Rule. As we noted in the previous unit, we can change the base of a logarithm to any other base that we want to. Most calculators have a *log* button and a *ln* button. We can use either of these to evaluate any log that we wish. Let's do an example.

Example 9: Use the Change of Base Rule to evaluate $\log_8 32$.

We use the rule to rewrite $\log_8 32$ as $\dfrac{\ln 32}{\ln 8}$. Now if we use our calculator, we get: $\dfrac{\ln 32}{\ln 8} = \dfrac{5}{3}$. Note that this is the same answer that we got in Unit 6, Example 7.

It would have been a big surprise if we had gotten a different answer!

Why would we want to use natural logs rather than common logs? As we said earlier, many natural phenomena change exponentially and, when analyzed, the solution often involves e, so we would need natural logs to work with them. Also, in Calculus, the properties of e lead to many simple solutions to problems. Generally, these problems are much more complicated to solve if they are evaluated using a log with a different base than e. So get used to natural logs and e. We will be using them a lot!

Are you ready to practice?

Practice Problem Set #7

1) If you invest $5000 in the bank at 8% interest, compounded annually, how much will you have after five years?

2) If you invest $10,000 in the bank at 4% interest, compounded semi-annually, how much will you have after ten years?

3) If you invest $8000 in the bank at 12% interest, compounded quarterly, how much will you have after 20 years?

4) If you invest $3000 in the bank at 6% interest, compounded monthly, how much will you have after three years?

5) If you invest $1000 in the bank at 10% interest, compounded weekly, how much will you have after seven years?

6) If you invest $4000 in the bank at 5% interest, compounded continuously, how much will you have after five years?

Expand the following logarithms:

7) $\ln \dfrac{5}{x}$

8) $\ln(3ab)$

9) $\ln\left(\dfrac{10A}{C}\right)$

10) $\ln\left(x^2 y^5\right)$

11) $\ln\left(\dfrac{x\sqrt[4]{y}}{z^6}\right)$

12) $\ln\sqrt{\dfrac{a^3}{b^2}}$

Write the following as a single logarithm:

13) $\ln 11 - \ln 2$

14) $3\ln A + 4\ln B - 8\ln C$

15) $6\ln w - \dfrac{1}{2}\ln z + 4\ln x$

16) $\dfrac{1}{3}\left(\ln x + 2\ln y - 5\ln z\right)$

17) $7\ln\left(xy^3\right) - 4\ln\left(y^5 z\right)$

18) $2\ln\left(x^2 y^3\right) + 6\ln(yz)$

Use the Change of Base Rule and your calculator to evaluate the following, using natural logarithms:

19) $\log_6 32$

20) $\log_7 9$

21) $\log_2 50$

Solutions to Practice Problem Set #7

1)   *If you invest $5000 in the bank at 8% interest, compounded annually, how much will you have after five years?*

Let's use our formula $F = P(1+R)^T$. Here, P is 5000, R is 0.08, and T is 5. Plugging in, we get: $F = 5000(1+0.08)^5 = \$7346.64$.

2)   *If you invest $10,000 in the bank at 4% interest, compounded semi-annually, how much will you have after ten years?*

Let's use our formula $F = P\left(1+\dfrac{R}{N}\right)^{NT}$. Here, P is 10,000, R is 0.04, N is 2, and T is 10. Plugging in, we get: $F = 10,000\left(1+\dfrac{0.04}{2}\right)^{2\cdot10} = 10000(1.02)^{20} = \$14,859.47$.

3)   *If you invest $8000 in the bank at 12% interest, compounded quarterly, how much will you have after 20 years?*

Let's use our formula $F = P\left(1+\dfrac{R}{N}\right)^{NT}$. Here, P is 8000, R is 0.12, N is 4, and T is 20. Plugging in, we get: $F = 8000\left(1+\dfrac{0.12}{4}\right)^{4\cdot20} = 8000(1.03)^{80} = \$85,127.12$.

4)   *If you invest $3000 in the bank at 6% interest, compounded monthly, how much will you have after three years?*

Let's use our formula $F = P\left(1+\dfrac{R}{N}\right)^{NT}$. Here, P is 3000, R is 0.06, N is 12, and T is 3. Plugging in, we get: $F = 3000\left(1+\dfrac{0.06}{12}\right)^{12\cdot3} = 3000(1.005)^{36} = \$3590.04$.

5) *If you invest $1000 in the bank at 10% interest, compounded daily how much will you have after seven years?*

Let's use our formula $F = P\left(1 + \dfrac{R}{N}\right)^{NT}$. Here, P is 1000, R is 0.10, N is 52, and T is 7. Plugging in, we get: $F = 1000\left(1 + \dfrac{0.10}{52}\right)^{52 \cdot 7} = \$2012.40$.

6) *If you invest $4000 in the bank at 5% interest, compounded continuously, how much will you have after five years?*

Now we are working with interest that compounds continuously, so we use our formula $F = Pe^{RT}$. Here, P is 4000, R is 0.05, and T is 5. Plugging in, we get: $F = 4000e^{0.05 \cdot 5} = \$5136.10$.

Expand the following logarithms:

7) $\ln\dfrac{5}{x}$

Here, we can use the Quotient Law, which says that $\ln\dfrac{A}{B} = \ln A - \ln B$.
We get: $\ln\dfrac{5}{x} = \ln 5 - \ln x$.

8) $\ln(3ab)$

Here, we can use the Product Law, which says that $\ln(AB) = \ln A + \ln B$. Of course, if we have more than two terms multiplied together, we can add all of their logs.

We get: $\ln(3ab) = \ln 3 + \ln a + \ln b$.

9) $\ln\left(\dfrac{10A}{C}\right)$

First, we can use the Quotient Law: $\ln\dfrac{10A}{C} = \ln 10A - \ln C$. Next, we use the Product Law to get: $\ln(10A) - \ln C = \ln 10 + \ln A - \ln C$.

10) $\ln\left(x^2 y^5\right)$

First, we can use the Product Law: $\ln\left(x^2 y^5\right) = \ln x^2 + \ln y^5$. Next, we can use the Power Law, which says that $\ln\left(A^B\right) = B\ln A$. We get: $\ln\left(x^2\right) + \ln\left(y^5\right) = 2\ln x + 5\ln y$.

11) $\ln\left(\dfrac{x\sqrt[4]{y}}{z^6}\right)$

First, we can use the Quotient Law: $\ln\left(\dfrac{x\sqrt[4]{y}}{z^6}\right) = \ln\left(x\sqrt[4]{y}\right) - \ln\left(z^6\right)$. Next,

we can use the Product Law: $\ln\left(x\sqrt[4]{y}\right) - \ln\left(z^6\right) = \ln x + \ln\sqrt[4]{y} - \ln\left(z^6\right)$. Finally,

we can use the Power Law: $\ln x + \ln\sqrt[4]{y} - \ln\left(z^6\right) = \ln x + \dfrac{1}{4}\ln y - 6\ln z$.

12) $\ln\sqrt{\dfrac{a^3}{b^2}}$

First, let's rewrite the radical as a power: $\ln\sqrt{\dfrac{a^3}{b^2}} = \ln\left[\left(\dfrac{a^3}{b^2}\right)^{\frac{1}{2}}\right]$. Now, let's

use the Power Law: $\ln\left[\left(\dfrac{a^3}{b^2}\right)^{\frac{1}{2}}\right] = \dfrac{1}{2}\ln\left(\dfrac{a^3}{b^2}\right)$. Next, let's use the Quotient Law:

$\dfrac{1}{2}\ln\left(\dfrac{a^3}{b^2}\right) = \dfrac{1}{2}\left[\ln\left(a^3\right) - \ln\left(b^2\right)\right]$. Finally, let's use the Power Law again to get:

$\dfrac{1}{2}\left[\ln\left(a^3\right) - \ln\left(b^2\right)\right] = \dfrac{1}{2}(3\ln a - 2\ln b)$.

Write the following as a single logarithm:

13) $\ln 11 - \ln 2$

Here, we can use the Quotient Law, which says that $\ln\dfrac{A}{B} = \ln A - \ln B$.

We get: $\ln 11 - \ln 2 = \ln\dfrac{11}{2}$.

14) $3\ln A + 4\ln B - 8\ln C$

First, we can use the Power Law, which says that $\ln\left(A^B\right) = B\ln A$. We get:

$3\ln A + 4\ln B - 8\ln C = \ln\left(A^3\right) + \ln\left(B^4\right) - \ln\left(C^8\right)$. Next, we can use the Product

Law, which says that $\ln AB = \ln A + \ln B$, to get: $\ln\left(A^3\right) + \ln\left(B^4\right) - \ln\left(C^8\right) =$

$\ln\left(A^3 B^4\right) - \ln\left(C^8\right)$. Finally, we can use the Quotient Law: $\ln\left(A^3 B^4\right) - \ln\left(C^8\right) =$

$\ln\left(\dfrac{A^3 B^4}{C^8}\right)$.

15) $6\ln w - \dfrac{1}{2}\ln z + 4\ln x$

First, we can use the Power Law: $6\ln w - \dfrac{1}{2}\ln z + 4\ln x = \ln w^6 - \ln z^{\frac{1}{2}} + \ln x^4$.

Next, we can use the Product Law to get: $\ln w^6 - \ln z^{\frac{1}{2}} + \ln x^4 = \ln w^6 x^4 - \ln z^{\frac{1}{2}}$.
Notice that we combined the first and the third logs. We don't have to take
them in order! Finally, we can use the Quotient Law: $\ln w^6 x^4 - \ln z^{\frac{1}{2}} =$

$\ln\left(\dfrac{w^6 x^4}{z^{\frac{1}{2}}}\right)$, or $\ln\left(\dfrac{w^6 x^4}{\sqrt{z}}\right)$.

16) $\dfrac{1}{3}\left(\ln x + 2\ln y - 5\ln z\right)$

First, we can use the Power Law for the terms inside the parentheses.
We will leave the term outside the parentheses for last. We get:
$\dfrac{1}{3}\left(\ln x + 2\ln y - 5\ln z\right) = \dfrac{1}{3}\left[\ln x + \ln\left(y^2\right) - \ln\left(z^5\right)\right]$. Next, we can use the

Product Law to get: $\dfrac{1}{3}\left[\ln x + \ln\left(y^2\right) - \ln\left(z^5\right)\right] = \dfrac{1}{3}\left[\ln\left(xy^2\right) - \ln\left(z^5\right)\right]$. Now, we

can use the Quotient Law to get: $\dfrac{1}{3}\left[\ln\left(xy^2\right) - \ln\left(z^5\right)\right] = \dfrac{1}{3}\left[\ln\left(\dfrac{xy^2}{z^5}\right)\right]$. Finally,

we can use the Power Law again to get: $\dfrac{1}{3}\left(\ln\dfrac{xy^2}{z^5}\right) = \ln\left(\dfrac{xy^2}{z^5}\right)^{\frac{1}{3}}$, or $\ln\sqrt[3]{\left(\dfrac{xy^2}{z^5}\right)}$.

17) $7\ln\left(xy^3\right) - 4\ln\left(y^5 z\right)$

First, we can use the Power Law on the two logs separately:
$7\ln\left(xy^3\right) - 4\ln\left(y^5 z\right) = \ln\left[\left(xy^3\right)^7\right] - \ln\left[\left(y^5 z\right)^4\right]$. Next, we can use the Quotient

Law to get: $\ln\left[\left(xy^3\right)^7\right] - \ln\left[\left(y^5 z\right)^4\right] = \ln\left[\dfrac{\left(xy^3\right)^7}{\left(y^5 z\right)^4}\right]$. Although we could stop

here, let's simplify the logarithm. Distribute the powers to their terms:
$\ln\left[\dfrac{\left(xy^3\right)^7}{\left(y^5 z\right)^4}\right] = \ln\left(\dfrac{x^7 y^{21}}{y^{20} z^4}\right)$. Then combine like terms: $\ln\left(\dfrac{x^7 y^{21}}{y^{20} z^4}\right) = \ln\left(\dfrac{x^7 y}{z^4}\right)$.

18) $2\ln(x^2y^3)+6\ln(yz)$

First, we can use the Power Law on the two logs separately: $2\ln(x^2y^3)+6\ln(yz)=\ln\left[(x^2y^3)^2\right]+\ln\left[(yz)^6\right]$. Next, we can use the Product Law to get: $\ln(x^2y^3)^2+\ln(yz)^6=\ln\left[(x^2y^3)^2(yz)^6\right]$. Although we could stop here, let's simplify the logarithm. Distribute the powers to their terms: $\ln\left[(x^2y^3)^2(yz)^6\right]=\ln\left[(x^4y^6)(y^6z^6)\right]$. Then combine like terms: $\ln\left[(x^4y^6)(y^6z^6)\right]=\ln(x^4y^{12}z^6)$.

Use the Change of Base Rule and your calculator to evaluate the following, using natural logarithms:

19) $\log_6 32$

The Change of Base Rule says that $\log_b x=\dfrac{\ln x}{\ln b}$. We can change the logarithms to any base we want, but if we change the base to natural logs, we will be able to use a calculator. We get: $\log_6 32=\dfrac{\ln 32}{\ln 6}\approx\dfrac{3.4657}{1.7918}\approx 1.9342$. Note that this is the same answer that we got in Unit Six.

20) $\log_7 9$

The Change of Base Rule says: $\log_b x=\dfrac{\ln x}{\ln b}$. We can change the logarithms to any base we want, but if we change the base to natural logs, we will be able to use a calculator. We get: $\log_7 9=\dfrac{\ln 9}{\ln 7}\approx\dfrac{2.1972}{1.9459}\approx 1.1291$. Note that this is the same answer that we got in Unit Six.

21) $\log_2 50$

The Change of Base Rule says that $\log_b x=\dfrac{\ln x}{\ln b}$. We can change the logarithms to any base we want, but if we change the base of the logarithms to base 10, we will be able to use a calculator. We get: $\log_2 50=\dfrac{\ln 50}{\ln 2}\approx\dfrac{3.9120}{0.6931}\approx 5.6442$. Note that this is the same answer that we got in Unit Six.

UNIT EIGHT

Graphs of Logarithmic Functions

Now that we are comfortable with logarithms, let's look at logarithmic functions. These show up in many "real-world" situations, such as compound interest, radioactive decay, population growth, and more, and are often linked to exponential functions. Here, we will look at the graphs of logarithmic functions, which is an excellent way to visualize logarithmic behavior.

First, let's remember how an exponential function behaves. For example, let's look at the function $y = 10^x$ and make a table of values:

x	y
0	1
1	10
2	100
3	1000
4	10,000

Notice how rapidly y grows. When x is 0, y is 1, and when x is 4, y is 10,000.

The graph of $y = 10^x$ looks like this:

Figure 1

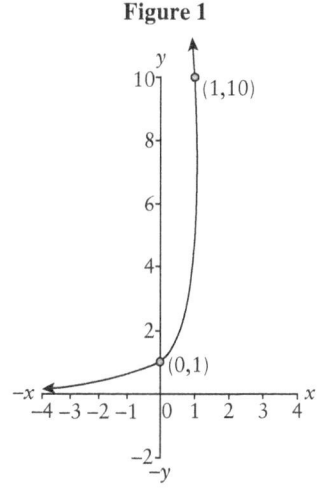

Here is what the graph looks like:

Figure 4

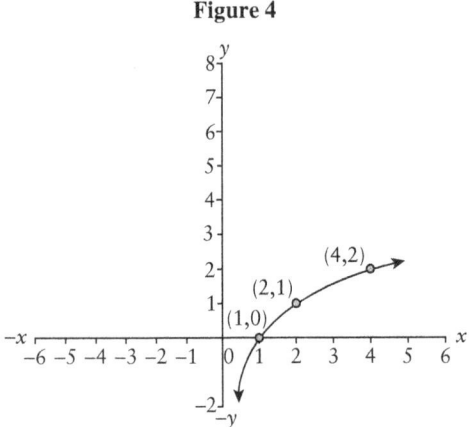

Notice the similarity in shape to the graph of $y = \log x$. Notice also that we only have values for positive values of x, and that y can have any value. That is, the domain is $x > 0$ and the range is $-\infty < y < \infty$. Remember that the domain of an exponential function is $-\infty < x < \infty$ and the range is $y > 0$. We expected this because the logarithmic function and the exponential function are inverses.

Let's note some of the key features of a log graph. If we are graphing $y = \log_b x$, it has a vertical asymptote at the y-axis $(x = 0)$, it has an x-intercept at $(1,0)$, and it goes through the point $(b,1)$.

Let's do another example.

Example 2: Now let's look at the graph of $y = \log_4 x$.

Figure 5

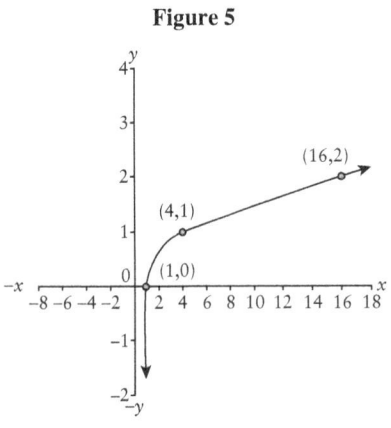

Note that the graph has the same shape as the other graphs. It has the same vertical asymptote and x-intercept, and it goes through the point $(4,1)$. Now let's see some shifts.

Example 3: Let's look at the graph of $y = \log_2(x-1)$.

Figure 6

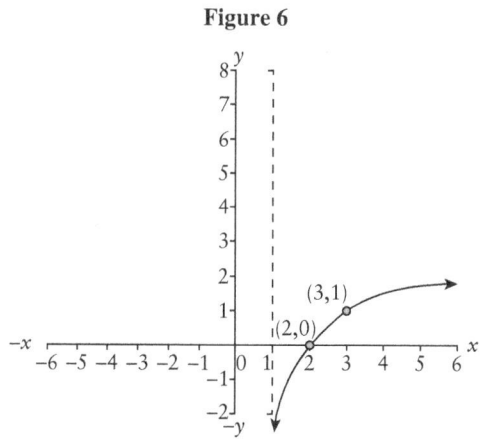

Here, the graph looks similar to $y = \log_2 x$, but it is shifted one unit to the right. So, its vertical asymptote is the line $x = 1$, its x-intercept is at $(2,0)$, and it goes through the point $(3,1)$.

Example 4: Finally, let's look at the graph of $y = \log_2(x-1)+3$.

Figure 7

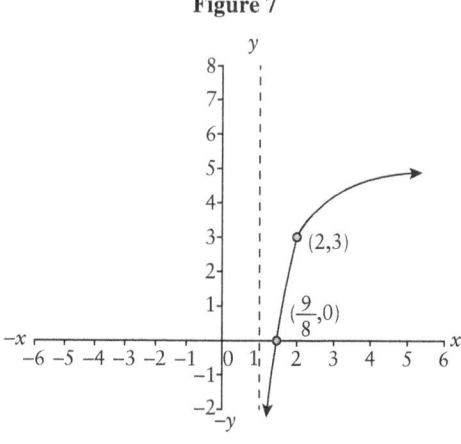

Here the graph again looks similar to $y = \log_2 x$, but is shifted one unit to the right and goes through $(2,3)$. It is as if the x-intercept has been shifted up by three units. This is because $y = \log_2(2-1) = 0$ at $x = 2$. So, instead of noting the x-intercept, we note the point that it has been shifted to. If we want to find the x-intercept, we need to figure out the value of x that will make $\log_2(x-1) = -3$. We can solve for x and get $x = 2^{-3} + 1 = \dfrac{9}{8}$. Therefore, the *x-intercept* is $\left(\dfrac{9}{8}, 0\right)$

Note all of the similarities to the exponential functions. This is one of the most important aspects of logarithmic and exponential functions—because they are inverses, they mirror each other.

Let's make a rule for graphing logarithmic functions.

If we have a function of the form $y = \log_b(x-c) + d$ where $b > 1$, c is the vertical asymptote, the x intercept is at $\left(b^{-d} + c, 0\right)$ and the graph goes through the point $(c+1, d)$.

Figure 8

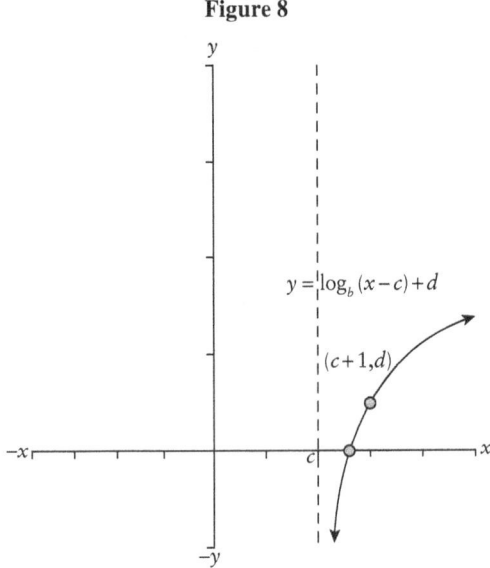

Now, let's look at a graph where $0 < b < 1$.

Example 5: Graph $y = \log_{\frac{1}{2}} x$. First, let's make a table of values:

x	y
8	-3
4	-2
2	-1
1	0

$\dfrac{1}{2}$	1
$\dfrac{1}{4}$	2

Notice that the same x values occur as we had for the graph $y = \log_2 x$, but now they yield negative y-values instead of positive ones. This will have the effect of reflecting the graph of $y = \log_2 x$ across the x axis.

Figure 9

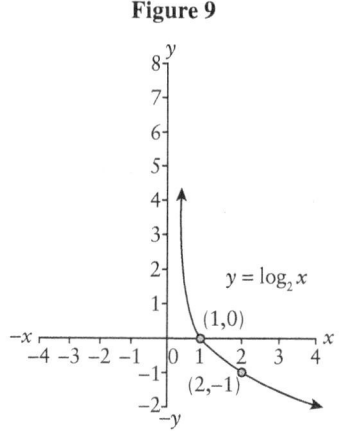

Note that the graph of $y = \log_{\frac{1}{2}} x$ has the same x-intercept as the graph of $y = \log_2 x$, the same vertical asymptote, and the same shape. Let's graph them together so we can compare them:

Figure 10

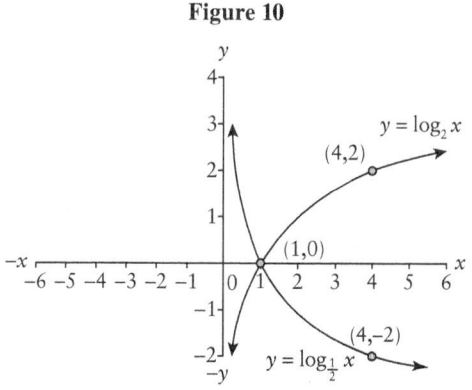

Let's look at the graphs of $y = \ln x$ and $y = e^x$. Of course, the graphs will look the same as the other logarithmic and exponential graphs.

Example 6: Let's look at $y = \ln x$.
  Let's make a table of values:

x	y
1	0
e	1
e^2	2
e^3	3
e^4	4

Note that they follow the same pattern as the other logarithmic graphs, but that the x values are in terms of e. The graph looks like this:

Figure 11

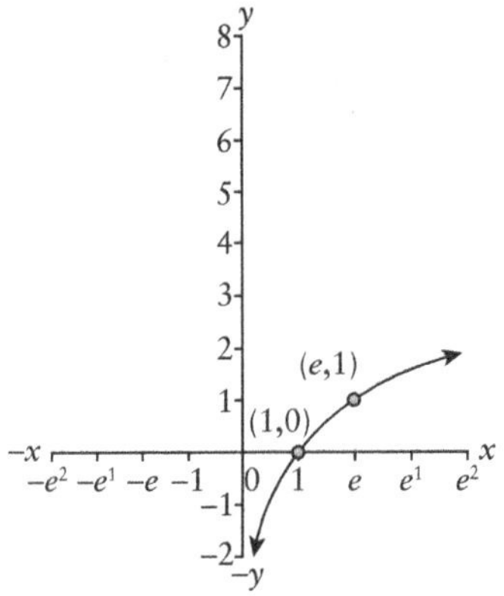

It has the same vertical asymptote, the y-axis, the same x-intercept $(1,0)$, and so on. Don't be intimidated by the graph of $y = \ln x$; it is just like the other log graphs.

Example 6: Finally, let's look at $y = e^x$.
  Let's make a table of values.

x	y
0	1
1	e

2	e^2
3	e^3
4	e^4

Note that they follow the same pattern as the other exponential graphs, but that the y values are in terms of e. The graph looks like this:

Figure 12

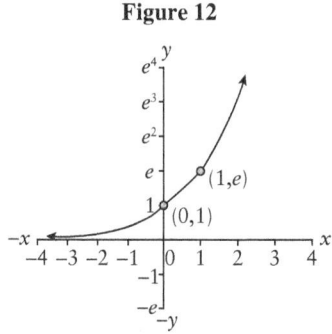

It has the same horizontal asymptote, the x-axis, the same y-intercept $(0,1)$, and so on. So again, don't be intimidated by the graph of $y = e^x$; it is just like the other exponential graphs.

Now we have the basics of logarithmic graphs. Are you ready for some practice problems?

Practice Problem Set #8

Find the vertical asymptote, the x-intercept, and one other point, and graph each of the following functions:

1) $y = \log_6 x$

2) $y = \ln(x+2)$

3) $y = \log_6(x-2)$

4) $y = \log x + 1$

5) $y = \log_4(x-1) + 2$

6) $y = \log_2(x-2) - 3$

7) $y = \log_{\frac{1}{3}}(x-1)$

8) $y = \log_{\frac{1}{4}}(x-2) - 1$

Find the horizontal asymptote, the y-intercept, and one other point. Graph each of the following functions:

9) $y = e^{x-3}$

10) $y = 3 \cdot e^{x+2} + 1$

Solutions to Practice Problem Set #8

Find the vertical asymptote, the x-intercept, and one other point, and graph each of the following functions:

1) $y = \log_6 x$

The x-intercept is $(1,0)$, it goes through $(6,1)$, and the vertical asymptote is $x = 0$. The graph looks like this:

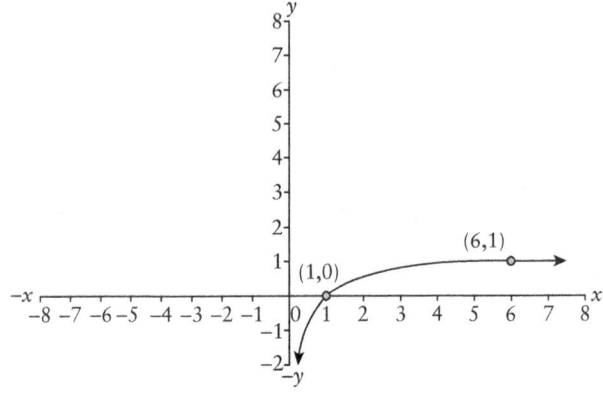

2) $y = \ln(x+2)$

The x-intercept is $(-1,0)$, it goes through $(e-2,1)$, and the vertical asymptote is $x = -2$. The graph looks like this:

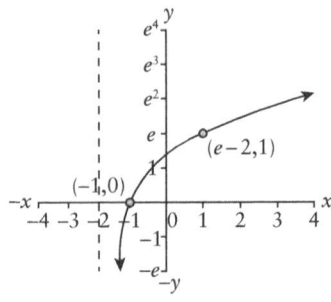

3) $y = \log_6(x - 2)$

The x-intercept is $(3,0)$, it goes through $(8,1)$, and the vertical asymptote is $x = 2$. The graph looks like this:

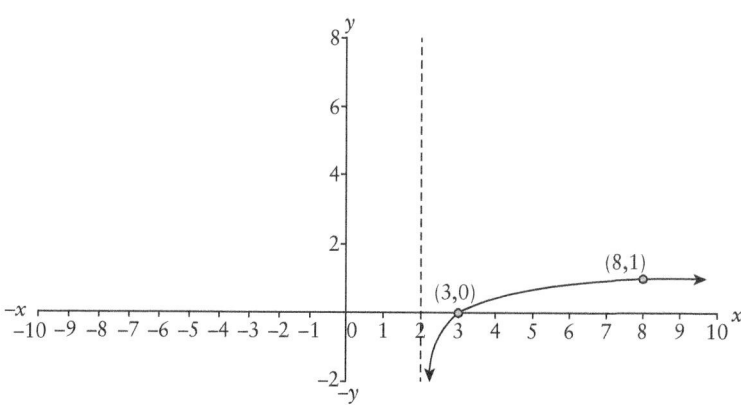

4) $y = \log x + 1$

The x-intercept is $\left(\dfrac{1}{10}, 0\right)$, it goes through $(1,1)$, and the vertical asymptote is $x = 0$. The graph looks like this:

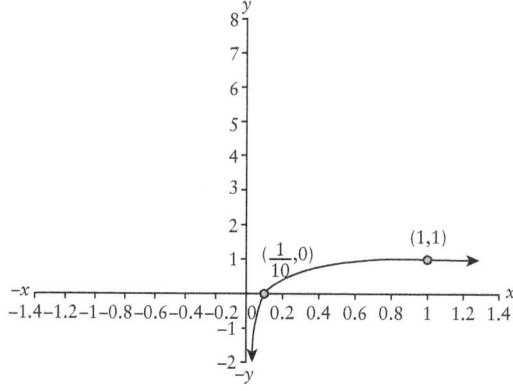

5) $y = \log_4(x - 1) + 2$

The x-intercept is $\left(\dfrac{1}{16} + 1, 0\right) = \left(\dfrac{17}{16}, 0\right)$, it goes through $(5,3)$, and the vertical asymptote is $x = 1$. The graph looks like this:

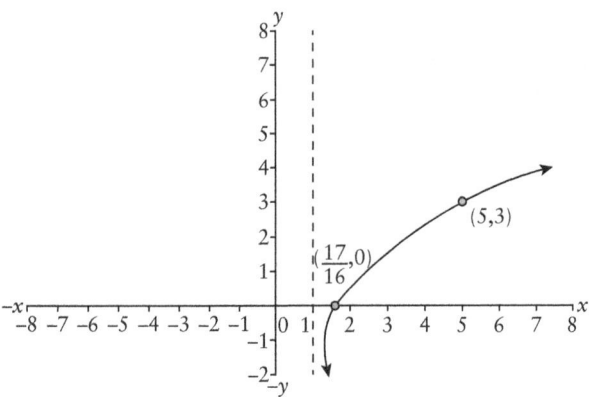

6) $y = \log_2(x-2) - 3$

The x-intercept is $(10,0)$, it goes through $(3,-3)$, and the vertical asymptote is $x = 2$. The graph looks like this:

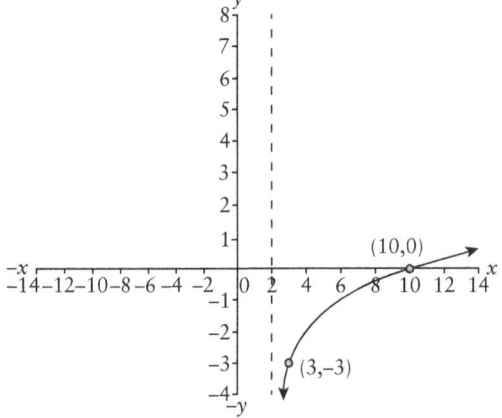

7) $y = \log_{\frac{1}{3}}(x-1)$

The x-intercept is $(2,0)$, it goes through $\left(\dfrac{4}{3},1\right)$, and the vertical asymptote is $x = 1$. The graph looks like this:

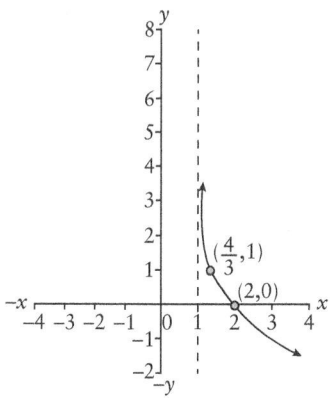

8) $y = \log_{\frac{1}{4}} (x - 2) - 1$

The x intercept is $\left(\dfrac{9}{4}, 0\right)$, it goes through $(3, -1)$, and the vertical asymptote is

$x = 2$. The graph looks like this:

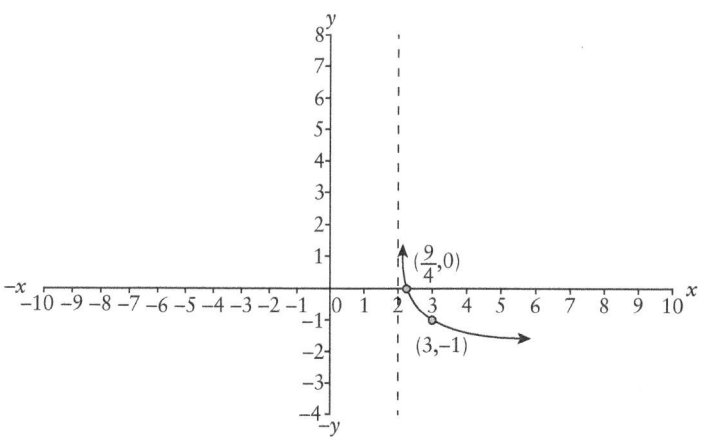

Find the horizontal asymptote, the y intercept, and one other point, and graph each of the following functions:

9) $y = e^{x-3}$

The y-intercept is $\left(0, \dfrac{1}{e^3}\right)$, which is about $(0, 0.5)$, it goes through $(3,1)$, and the horizontal asymptote is $y = 0$. The graph looks like this:

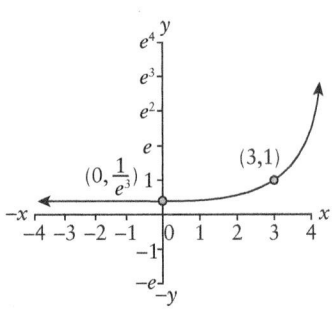

10) $y = 3 \cdot e^{x+2} + 1$

The y-intercept is $\left(0, 3e^2 + 1\right)$, which is about $(0, 23)$, it goes through $(-2, 4)$, and the horizontal asymptote is $y = 1$. The graph looks like this:

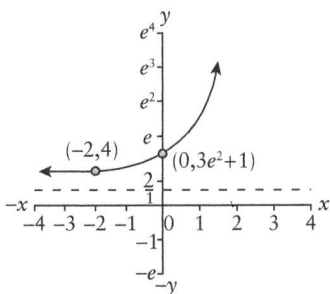

UNIT NINE

Problems That Use Exponentials

Often, we will encounter problems where a variable appears in the exponent or as a part of an exponential expression. An equation that uses such terms is called an exponential equation. First, let's learn how to solve an equation of the form $a \cdot b^x = c$, where a, b, and c are constants. Let's do an example.

Example 1: Solve for x: $27^{x-6} = 81$.

In order to solve this, we need to find a way to write both 27 and 81 in terms of the same base. Then, we can set their respective exponents equal to each other and solve. Here, we can write 27 as 3^3 and 81 as 3^4. Now we can rewrite the equation as $\left(3^3\right)^{x-6} = 3^4$. Remember that when we have an exponentiated expression raised to a power, we multiply the powers (see Unit One). Here we get: $3^{3x-18} = 3^4$. Next, because both expressions are in terms of the same base, 3, we can set the powers equal to each other: $3x - 18 = 4$. Now we can solve for x: $x = \dfrac{22}{3}$.

Let's do another example.

Example 2: Solve for x: $25^{3x+1} = 125$.

We need to write both 25 and 125 in terms of the same base. Here, we can write 25 as 5^2 and 125 as 5^3. Now we can rewrite the equation as $\left(5^2\right)^{3x+1} = 5^3$. We now can multiply the powers to get: $5^{6x+2} = 5^3$. Next, because both expressions are in terms of the same base, 5, we can set the powers equal to each other: $6x + 2 = 3$. We can solve for x to get $x = \dfrac{1}{6}$.

Now let's look at an equation where both sides are exponentiated.

Example 3: Solve for x: $8^{2x-5} = 4^{x+5}$.

We need to write both 8 and 4 in terms of the same base. We can write 8 as 2^3 and 4 as 2^2. Now we can rewrite the equation as $\left(2^3\right)^{2x-5} = \left(2^2\right)^{x+5}$. We now can multiply the powers to get: $2^{6x-15} = 2^{2x+10}$. Next, because both expressions are in terms of the same base, 2, we can set the powers equal to each other: $6x - 15 = 2x + 10$. We can solve for x to get $x = \dfrac{25}{4}$.

Note that this is essentially the same technique as the first example, except that both sides have a variable in the power instead of just one of the sides. Let's do another example.

Example 4: Solve for x: $16^{x+1} = 64^{x-2}$.

We need to write both 16 and 64 in terms of the same base. We actually have two options. We could rewrite them either as powers of 2 or powers of 4. Let's solve the equation both ways to see that the answers are the same either way. First, let's write 16 as 2^4 and 64 as 2^6. Now we can rewrite the equation as $\left(2^4\right)^{x+1} = \left(2^6\right)^{x-2}$. We now can multiply the powers to get: $2^{4x+4} = 2^{6x-12}$. Next, we can set the powers equal to each other: $4x + 4 = 6x - 12$. We can solve for x to get $x = 8$.

Now, let's redo the problem using 16 as 4^2 and 64 as 4^3. We multiply the powers to get: $4^{2x+2} = 4^{3x-6}$. Set the powers equal to each other: $2x + 2 = 3x - 6$. And solve for x: $x = 8$. So, we can see that it doesn't matter which base we use; either way, we get the same answer. If we didn't, we would have a big problem!

Suppose that we want to solve an equation like $5^x = 25$. This is easy to solve: the answer is $x = 2$. But how do we solve an equation like $5^x = 20$. We use logarithms!

Example 5: Solve for x: $5^x = 20$.

We cannot write 5 and 20 as powers of the same base, so we will use logarithms to solve the equation. Just as we can square both sides of an equation, we can take the log of both sides of an equation. By the way, it doesn't matter which base we use for the log, as long as it is the same for both sides. It is convenient to use either the common log or the natural log because we can use our calculators easily. In general, we will take the common log of both sides. Here we get: $\log 5^x = \log 20$. Now, we can use the Power Law to rewrite this as: $x \log 5 = \log 20$. Now we divide both sides by $\log 5$ to get: $x = \dfrac{\log 20}{\log 5} \approx 1.86$. Note that this is close to 2, so the answer is reasonable. When we solve an equation using logs, we should always check to see if the answer makes sense. Because $5^2 = 25$, we should expect a power of 5 that is a little less than 2.

One other point: suppose we had used natural logs instead of common logs. Then we would have gotten $x = \dfrac{\ln 20}{\ln 5} \approx 1.86$. Verify this with your calculator. So, as we noted before, it doesn't matter which log we use, as long as it is the same one for both sides and that we are consistent throughout the problem. Let's do another example.

Example 6: Solve for x: $4^{x-2} = 70$.

We can't write 4 and 70 in terms of the same base, so we take the log of both sides. We get: $\log 4^{x-2} = \log 70$. Next, bring the power in front of the log to get $(x-2)\log 4 = \log 70$. Now we distribute the $\log 4$ on the left side: $x \log 4 - 2 \log 4 = \log 70$. Add $2 \log 4$ to both sides: $x \log 4 = \log 70 + 2 \log 4$. And divide both sides by $\log 4$: $x = \dfrac{\log 70 + 2 \log 4}{\log 4} \approx 5.065$. We were expecting an answer around 5 because $4^{5-2} = 4^3 = 64$, and 70 is close to 64. This wasn't very difficult but we can see that it is easy to mess up the Algebra, so be careful! Let's do a more complicated example, where both sides have a variable in the exponent.

Example 7: Solve for x: $3^{x+1} = 5^x$.

We can't write 3 and 5 in terms of the same base, so we take the log of both sides. We get: $\log 3^{x+1} = \log 5^x$. Next, bring the powers in front of the logs to get $(x+1)\log 3 = x\log 5$. Now we distribute the log terms: $x\log 3 + \log 3 = x\log 5$. Now what do we do? There are three steps to follow: **Group, Factor, Divide**.

First, we **Group** the terms that contain an x on one side and the terms that do not on the other side: $\log 3 = x\log 5 - x\log 3$.

Second, we **Factor** out the x: $\log 3 = x(\log 5 - \log 3)$.

Last, we **Divide** to isolate x: $\dfrac{\log 3}{\log 5 - \log 3} = x \approx 2.15$.

Remember this technique: Group, Factor, Divide. It is what we will use to solve many types of exponential equations. Let's do another one.

Example 8: Solve for x: $6^{2-x} = 4^{3x+1}$.

We can't write 6 and 4 in terms of the same base, so we take the log of both sides. We get: $\log 6^{2-x} = \log 4^{3x+1}$. Next, bring the powers in front of the logs to get $(2-x)\log 6 = (3x+1)\log 4$. Now we distribute the log terms: $2\log 6 - x\log 6 = 3x\log 4 + \log 4$.

Now we **Group, Factor, Divide**.

First, we **Group** the terms that contain an x on one side and the terms that do not on the other side: $2\log 6 - \log 4 = 3x\log 4 + x\log 6$.

Second, we **Factor** out the x: $2\log 6 - \log 4 = x(3\log 4 + \log 6)$.

Last, we **Divide** to isolate x: $\dfrac{2\log 6 - \log 4}{3\log 4 + \log 6} = 0.369$.

Remember the finance problems that we did in Unit Seven? We can use logarithms to solve some of these types of problems. Let's do an example.

Example 9: If we deposit $5000 in a bank that pays 6% interest compounded annually, how long will it take for the money to double?

Remember that the compound interest formula is $F = P(1+R)^T$, where F is the future amount, P is what we invest now (the present amount), R is the interest rate (expressed as a decimal), and T is the number of years. We want to figure out how long we will need to leave our money in the bank for it to reach $10,000. Plugging in, we get: $10,000 = 5000(1+0.06)^T$. Now we divide both sides by 5000: $2 = (1.06)^T$. Notice that we are trying to solve for the power. We can do this by taking the log of both sides. We get: $\log 2 = \log(1.06)^T$. Next, we bring the power in front of the log: $\log 2 = T \log(1.06)$. Now, divide both sides by $\log 1.06$: $\dfrac{\log 2}{\log 1.06} = T \approx 11.90$ years.

Suppose we put our money in a bank that pays interest that compounds continuously?

Example 10: If we deposit \$5000 in a bank that pays 6% interest compounded continuously, how long will it take for the money to double?

Remember that the compound interest formula here is $F = Pe^{RT}$, where F is the future amount, P is what we invest now (the present amount), R is the interest rate (expressed as a decimal), and T is the number of years. We want to figure out how long we will need to leave our money in the bank for it to reach \$10,000. Plugging in, we get: $10,000 = 5000e^{0.06T}$. Now, we divide both sides by 5000: $2 = e^{0.06T}$.

Notice that here we are trying to solve for the power of e. We can do this by taking the log of both sides but, because we are using e let's take the natural log of both sides (it makes life a little easier doing it this way when we do the algebra). We get: $\ln 2 = \ln e^{0.06T}$. Next, we bring the power in front of the log: $\ln 2 = 0.06T \ln e$. Because $\ln e = 1$, we get: $\ln 2 = 0.06T$. Now we can divide both sides by 0.06: $\dfrac{\ln 2}{0.06} = T \approx 11.55$ years.

As we can see, logarithms are a very handy tool for solving problems where the power is an unknown variable. Are you ready for some practice problems?

Practice Problem Set #9

1)   Solve for x: $8^{2x+1} = 128$.

2)   Solve for x: $7^{4x+5} = 343$.

3)   Solve for x: $6^{5-4x} = 216$.

4)   Solve for x: $25^{2x+1} = 125^{3x-1}$.

5)   Solve for x: $81^{5x-2} = 729^{x+1}$.

6)   Solve for x: $1000^{1-4x} = 100^{3+2x}$.

7)   Solve for x: $27^{4-x} = 243^{x+1}$.

8)   Solve for x: $8^{x+3} = 30$.

9)   Solve for x: $5^{7-2x} = 44$.

10)   Solve for x: $4^{2-x} = 110$.

11)   Solve for x: $12^{x-7} = 88$.

12) Solve for x: $11^{x+7} = 4^x$.

13) Solve for x: $9^{2x+1} = 8^{x-3}$.

14) Solve for x: $7^{2-5x} = 5^{3x+4}$.

15) Solve for x: $2^{4x-9} = 10^{x+7}$.

16) Solve for x: $13^{9-5x} = 6^{x+2}$.

17) Solve for x: $8^{7-x} = 6^{3-8x}$.

18) If we deposit $1000 in a bank that pays 4% interest compounded annually, how long will it take until we have $3000?

19) If we deposit $2000 in a bank that pays 3% interest compounded monthly, how long will it take until we have $5000?

20) If we deposit $10,000 in a bank that pays 6% interest compounded quarterly, how long will it take to double?

21) If we deposit $1000 in a bank that pays 5% interest compounded continuously, how long will it take until we have $4000?

Solutions to Practice Problem Set #9

1) *Solve for x:* $8^{2x+1} = 128$.

In order to solve this, we need to find a way to write both 8 and 128 in terms of the same base. Then, we can set their respective exponents equal to each other and solve. Here, we can write 8 as 2^3 and 128 as 2^7. Now we can rewrite the equation as $\left(2^3\right)^{2x+1} = 2^7$. Remember that when we have an exponentiated expression raised to a power, we multiply the powers (see Unit One). Here we get: $2^{6x+3} = 2^7$. Next, because both expressions are in terms of the same base, 2, we can set the powers equal to each other: $6x + 3 = 7$. Now we can solve for x: $x = \dfrac{2}{3}$.

2) *Solve for x:* $7^{4x+5} = 343$.

In order to solve this, we need to find a way to write both 7 and 343 in terms of the same base. Then, we can set their respective exponents equal to each other and solve. Here, we can write 343 as 7^3. Now we can rewrite the equation as $7^{4x+5} = 7^3$. Now, because both expressions are in terms of the same base, 7, we can set the powers equal to each other: $4x + 5 = 3$. Now we can solve for x: $x = -\dfrac{1}{2}$.

3) *Solve for x:* $6^{5-4x} = 216$.

In order to solve this, we need to find a way to write both 6 and 216 in terms of the same base. Then, we can set their respective exponents equal to each other and solve. Here, we can write 216 as 6^3. Now we can rewrite the equation as $6^{5-4x} = 6^3$. Now, because both expressions are in terms of the same base, 6, we can set the powers equal to each other: $5 - 4x = 3$. Now we can solve for x: $x = \dfrac{1}{2}$.

4) *Solve for x:* $25^{2x+1} = 125^{3x-1}$.

In order to solve this, we need to find a way to write both 25 and 125 in terms of the same base. Then, we can set their respective exponents equal to each other and solve. Here, we can write 25 as 5^2 and 125 as 5^3. Now we can rewrite the equation as $\left(5^2\right)^{2x+1} = \left(5^3\right)^{3x-1}$. Remember that when we have an exponentiated expression raised to a power, we multiply the powers (see Unit One). Here, we get: $5^{4x+2} = 5^{9x-3}$. Now, because both expressions are in terms of the same base, 5, we can set the powers equal to each other: $4x + 2 = 9x - 3$. Now we can solve for x: $x = 1$.

5) *Solve for x:* $81^{5x-2} = 729^{x+1}$.

In order to solve this, we need to find a way to write both 81 and 729 in terms of the same base. Then, we can set their respective exponents equal to each other and solve. Here, we can write 81 as 9^2 and 729 as 9^3. Now we can rewrite the equation as $\left(9^2\right)^{5x-2} = \left(9^3\right)^{x+1}$. Remember that when we have an exponentiated expression raised to a power, we multiply the powers (see Unit One). Here, we get: $9^{10x-4} = 9^{3x+3}$. Now, because both expressions are in terms of the same base, 9, we can set the powers equal to each other: $10x - 4 = 3x + 3$. Now we can solve for x: $x = 1$. By the way, we could have chosen 3 as our base instead of 9. Then we would have written 81 as 3^4 and 729 as 3^6 and proceeded from there. Of course, we still would have obtained a solution of $x = 1$.

6) *Solve for x:* $1000^{1-4x} = 100^{3+2x}$.

In order to solve this, we need to find a way to write both 1000 and 100 in terms of the same base. Then, we can set their respective exponents equal to each other and solve. Here, we can write 1000 as 10^3 and 100 as 10^2. Now we can rewrite the equation as $\left(10^3\right)^{1-4x} = \left(10^2\right)^{3+2x}$. Remember that when we have an exponentiated expression raised to a power, we multiply the powers (see Unit One). Here, we get: $10^{3-12x} = 10^{6+4x}$. Now, because both expressions are in terms of the same base, 10, we can set the powers equal to each other: $3 - 12x = 6 + 4x$. Now we can solve for x: $x = -\dfrac{3}{16}$.

7) *Solve for x:* $27^{4-x} = 243^{x+1}$.

In order to solve this, we need to find a way to write both 27 and 243 in terms of the same base. Then, we can set their respective exponents equal to each other and solve. Here, we can write 27 as 3^3 and 243 as 3^5. Now we can rewrite the equation as $\left(3^3\right)^{4-x} = \left(3^5\right)^{x+1}$. Remember that when we have an exponentiated expression raised to a power, we multiply the powers (see Unit One). Here, we get: $3^{12-3x} = 3^{5x+5}$. Now, because both expressions are in terms of the same base, 3, we can set the powers equal to each other: $12 - 3x = 5x + 5$. Now we can solve for x: $x = \dfrac{7}{8}$.

8) *Solve for x:* $8^{x+3} = 30$.

We cannot write 8 and 30 as powers of the same base, so we will use logarithms to solve the equation. Here we get: $\log 8^{x+3} = \log 30$. Now, we can use the Power Law to rewrite this as: $(x+3)\log 8 = \log 30$. Now we divide both sides by $\log 8$ to get: $x + 3 = \dfrac{\log 30}{\log 8}$. Now we just have to subtract 3 from both sides: $x = \dfrac{\log 30}{\log 8} - 3 \approx -1.36$. Suppose we had used natural logs instead of common logs. Then we would have gotten $x = \dfrac{\ln 30}{\ln 8} - 3 \approx -1.36$. So, as we noted before, it doesn't matter which log we use, as long as it is the same one for both sides.

9) *Solve for x:* $5^{7-2x} = 44$.

We cannot write 5 and 44 as powers of the same base, so we will use logarithms to solve the equation. Here we get: $\log 5^{7-2x} = \log 44$. Now, we can use the Power Law to rewrite this as: $(7-2x)\log 5 = \log 44$. Now we divide both sides by $\log 5$ to get: $7 - 2x = \dfrac{\log 44}{\log 5}$. With a little algebra, we get: $x = \dfrac{7 - \dfrac{\log 44}{\log 5}}{2} \approx 2.32$.

10) *Solve for x:* $4^{2-x} = 110$.

We cannot write 4 and 110 as powers of the same base, so we will use logarithms to solve the equation. Here we get: $\log 4^{2-x} = \log 110$. Now, we can use the Power Law to rewrite this as: $(2-x)\log 4 = \log 110$. Now we divide both sides by $\log 4$ to get: $2 - x = \dfrac{\log 110}{\log 4}$. With a little algebra, we get: $x = 2 - \dfrac{\log 110}{\log 4} \approx -1.39$.

11) *Solve for x:* $12^{x-7} = 88$.

We cannot write 12 and 88 as powers of the same base, so we will use logarithms to solve the equation. Here we get: $\log 12^{x-7} = \log 88$. Now, we can use the Power Law to rewrite this as: $(x-7)\log 12 = \log 88$. Now we divide both sides by $\log 12$ to get: $x - 7 = \dfrac{\log 88}{\log 12}$. Add 7 to both sides: $x = 7 + \dfrac{\log 88}{\log 12} \approx 8.80$.

12) *Solve for x:* $11^{x+7} = 4^x$.

We can't write 11 and 4 in terms of the same base, so we take the log of both sides. We get: $\log 11^{x+7} = \log 4^x$. Next, bring the powers in front of the logs to get: $(x+7)\log 11 = x\log 4$. Now we distribute the log term on the left side: $x\log 11 + 7\log 11 = x\log 4$.

Now what do we do? There are three steps to follow: **Group, Factor, Divide**.

First, we **Group** the terms that contain an x on one side and the terms that do not on the other side: $7\log 11 = x\log 4 - x\log 11$.

Second, we **Factor** out the x: $7\log 11 = x(\log 4 - \log 11)$.

Last, we **Divide** to isolate x: $\dfrac{7\log 11}{\log 4 - \log 11} = x \approx -16.59$.

13) *Solve for x:* $9^{2x+1} = 8^{x-3}$.

We take the log of both sides: $\log 9^{2x+1} = \log 8^{x-3}$. Next, bring the powers in front of the logs to get $(2x+1)\log 9 = (x-3)\log 8$. We distribute the log term on each side: $2x\log 9 + \log 9 = x\log 8 - 3\log 8$.

Now we **Group, Factor, Divide**.

First, we **Group** the terms that contain an x on one side and the terms that do not on the other side: $2x\log 9 - x\log 8 = -\log 9 - 3\log 8$.

Second, we **Factor** out the x: $x(2\log 9 - \log 8) = -\log 9 - 3\log 8$.

Last, we **Divide** to isolate x: $x = \dfrac{-\log 9 - 3\log 8}{2\log 9 - \log 8} = \dfrac{\log 9 + 3\log 8}{\log 8 - 2\log 9} \approx -3.64$.

14) *Solve for x:* $7^{2-5x} = 5^{3x+4}$.

We take the log of both sides: $\log 7^{2-5x} = \log 5^{3x+4}$. Next, bring the powers in front of the logs to get $(2-5x)\log 7 = (3x+4)\log 5$. We distribute the log term on each side: $2\log 7 - 5x\log 7 = 3x\log 5 + 4\log 5$.

Now, we **Group, Factor, Divide**.

First, we **Group** the terms that contain an x on one side and the terms that do not on the other side: $2\log 7 - 4\log 5 = 3x\log 5 + 5x\log 7$.

Second, we **Factor** out the x: $2\log 7 - 4\log 5 = x(3\log 5 + 5\log 7)$.

Last, we **Divide** to isolate x: $x = \dfrac{2\log 7 - 4\log 5}{3\log 5 + 5\log 7} \approx -0.175$.

15) *Solve for x:* $2^{4x-9} = 10^{x+7}$.

We take the log of both sides: $\log 2^{4x-9} = \log 10^{x+7} = x + 7$. Notice here that, by using log base 10, we eliminated the log from the right-hand side. Next, bring the power in front of the log in the left-hand term to get $(4x-9)\log 2 = x + 7$. We distribute the log term on the left side: $4x\log 2 - 9\log 2 = x + 7$.

Now, we **Group, Factor, Divide**.

First, we **Group** the terms that contain an x on one side and the terms that do not on the other side: $4x\log 2 - x = 9\log 2 + 7$.

Second, we **Factor** out the x: $x(4\log 2 - 1) = 9\log 2 + 7$.

Last, we **Divide** to isolate x: $x = \dfrac{9\log 2 + 7}{4\log 2 - 1} \approx 47.57$.

16) *Solve for x:* $13^{9-5x} = 6^{x+2}$.

We take the log of both sides: $\log 13^{9-5x} = \log 6^{x+2}$. Next, bring the powers in front of the logs to get $(9 - 5x)\log 13 = (x + 2)\log 6$. We distribute the log term on each side: $9\log 13 - 5x\log 13 = x\log 6 + 2\log 6$.

Now, we **Group, Factor, Divide**.

First, we **Group** the terms that contain an x on one side and the terms that do not on the other side: $9\log 13 - 2\log 6 = x\log 6 + 5x\log 13$.

Second, we **Factor** out the x: $9\log 13 - 2\log 6 = x(\log 6 + 5\log 13)$.

Last, we **Divide** to isolate x: $x = \dfrac{9\log 13 - 2\log 6}{\log 6 + 5\log 13} \approx 1.33$.

17) *Solve for x:* $8^{7-x} = 6^{3-8x}$.

We take the log of both sides: $\log 8^{7-x} = \log 6^{3-8x}$. Next, bring the powers in front of the logs to get $(7 - x)\log 8 = (3 - 8x)\log 6$. We distribute the log term on each side: $7\log 8 - x\log 8 = 3\log 6 - 8x\log 6$.

Now we **Group, Factor, Divide**.

First, we **Group** the terms that contain an x on one side and the terms that do not on the other side: $7\log 8 - 3\log 6 = x\log 8 - 8x\log 6$.

Second, we **Factor** out the x: $7\log 8 - 3\log 6 = x(\log 8 - 8\log 6)$.

Last, we **Divide** to isolate x: $x = \dfrac{7\log 8 - 3\log 6}{\log 8 - 8\log 6} \approx -0.75$.

18) *If we deposit $1000 in a bank that pays 4% interest compounded annually, how long will it take until we have $3000?*

Remember that the compound interest formula is $F = P(1+R)^T$, where F is the future amount, P is what we invest now (the present amount), R is the interest rate (expressed as a decimal), and T is the number of years. We want to figure out how long we will need to leave our money in the bank for it to reach $3000. Plugging in, we get: $3000 = 1000(1 + 0.04)^T$.

Now we divide both sides by 1000: $3 = (1.04)^T$.
Notice that we are trying to solve for the power. We can do this by taking the log of both sides. We get: $\log 3 = \log(1.04)^T$.
Next, we bring the power in front of the log: $\log 3 = T \log(1.04)$.
Divide both sides by $\log 1.04$: $\dfrac{\log 3}{\log 1.04} = T \approx 28.011$ years.

19) *If we deposit $2000 in a bank that pays 3% interest compounded monthly, how long will it take until we have $5000?*

Remember that the compound interest formula is $F = P\left(1 + \dfrac{R}{N}\right)^{NT}$, where F is the future amount, P is what we invest now (the present amount), R is the interest rate (expressed as a decimal), N is the number of compounding periods per year, and T is the number of years. We want to figure out how long we will need to leave our money in the bank for it to reach $5000. Plugging in, we get: $5000 = 2000\left(1 + \dfrac{0.03}{12}\right)^{12T}$. Note that because we are compounding interest monthly, $N = 12$.

Now we divide both sides by 2000: $2.5 = (1.0025)^{12T}$.
Notice that we are trying to solve for the power. We can do this by taking the log of both sides. We get: $\log 2.5 = \log(1.0025)^{12T}$.
Next, we bring the power in front of the log: $\log 2.5 = 12T \log(1.0025)$.
Divide both sides by $12 \log 1.0025$: $\dfrac{\log 2.5}{12 \log 1.0025} = T \approx 30.58$ years.

20) *If we deposit $10,000 in a bank that pays 6% interest compounded quarterly, how long will it take to double?*

Remember that the compound interest formula is $F = P\left(1 + \dfrac{R}{N}\right)^{NT}$, where F is the future amount, P is what we invest now (the present amount), R is the interest rate (expressed as a decimal), N is the number of compounding periods per year, and T is the number of years. We have $10,000 and we want the money to double, so we want to figure out how long we will need to leave our money in the bank for it to reach $20,000. Plugging in, we get: $20,000 = 10,000\left(1 + \dfrac{0.06}{4}\right)^{4T}$. Note that because we are compounding interest quarterly, $N = 4$.

Now we divide both sides by 10,000: $2 = (1.015)^{4T}$.
Notice that we are trying to solve for the power. We can do this by taking the log of both sides. We get: $\log 2 = \log(1.015)^{4T}$.
Next, we bring the power in front of the log: $\log 2 = 4T \log(1.015)$.
Divide both sides by $4 \log 1.015$: $\dfrac{\log 2}{4 \log 1.015} = T \approx 11.64$ years.

21) *If we deposit $1000 in a bank that pays 5% interest compounded continuously, how long will it take until we have $4000?*

Remember that the compound interest formula here is $F = Pe^{RT}$, where F is the future amount, P is what we invest now (the present amount), R is the interest rate (expressed as a decimal), and T is the number of years. We want to figure out how long we will need to leave our money in the bank for it to reach $4,000. Plugging in, we get: $4000 = 1000e^{0.05T}$.

Now we divide both sides by 1000: $4 = e^{0.05T}$.

Notice that here we are trying to solve for the power of e. We can do this by taking the log of both sides. Because we are using e, let's take the natural log of both sides. We get: $\ln 4 = \ln e^{0.05T}$.

Next, we bring the power in front of the log: $\ln 4 = 0.05T \ln e$. Because $\ln e = 1$, we get: $\ln 4 = 0.05T$.

Now we can divide both sides by 0.05: $\dfrac{\ln 4}{0.05} = T \approx 27.73$ years.

UNIT TEN

Problems That Use Logarithms

Now that we have learned how to solve equations and problems that use exponentials, we are going to learn how to solve problems that use logarithms. As with the previous unit, we will look at a variety of types of problems and see what we need to do to get to the solution. We will also look at a couple of different types of word problems that are solved with logarithms.

First, let's look at an equation that uses logarithms. We are used to equations that involve a variable, or more than one variable. There are also equations where the variable is a function of x, such as with trigonometric functions, exponentials, or, in this case, logarithms. We will use the laws of logarithms to solve them. Let's do an example.

Example 1: Solve for x: $\log(x+3) - \log(x-2) = \log 6$.

Remember the Log Laws. The Quotient Law says that $\log\dfrac{A}{B} = \log A - \log B$.

We can use this to rewrite the left side of the equation as $\log\left(\dfrac{x+3}{x-2}\right) = \log 6$. Now, because the two logs are equal, we can ignore the log parts and solve the equation $\dfrac{x+3}{x-2} = 6$. Cross-multiply: $x+3 = 6(x-2)$.

Then solve: $x+3 = 6x-12$, so $x = 3$.

It is always a good idea to check our answer with log equations because logarithms are only defined for positive values. If we plug $x = 3$ into the left side of the equation, we get: $\log(3+3) - \log(3-2)$, which reduces to: $\log 6 - \log 1$. Because $\log 1 = 0$, this becomes $\log 6$. So, our solution works.

That wasn't so bad! Let's try a similar one.

Example 2: Solve for x: $\log(x+3) + \log(x-2) = \log 6$.

Again, remember the Log Laws. The Product Law says that $\log AB = \log A + \log B$. We can use this to rewrite the left side of the equation as $\log[(x+3)(x-2)] = \log 6$. Now, because the two logs are equal, we can ignore the log parts and solve the equation $(x+3)(x-2) = 6$. This is a simple quadratic equation. Expand the left side: $x^2 + x - 6 = 6$. Subtract 6 from both sides: $x^2 + x - 12 = 0$. Now we can factor this and solve: $(x+4)(x-3) = 0$, so $x = -4$ or $x = 3$.

Let's check our answers. First, let's plug $x = -4$ into the left side of the equation. Note that we would then be taking the log of a negative number, which is not allowed. Thus, $x = -4$ is not a valid solution. Next, let's plug $x = 3$ into the left side of the equation: $\log(3+3) + \log(3-2)$. Which reduces to: $\log 6 + \log 1$. Just like last

time, because $\log 1 = 0$, this becomes $\log 6$. So, our second solution works. Therefore, even though we found two values of x, the only solution is $x = 3$.
Let's do another one.

Example 3: Solve for x: $\ln(x-2)+\ln(x+4)= \ln 7$.
The Product Law says that $\log(AB)= \log A + \log B$, so we can rewrite the left side of the equation as $\ln[(x-2)(x+4)] = \ln 7$. Now, because the two logs are equal, we can ignore the log parts and solve the equation $(x-2)(x+4)= 7$. This is a simple quadratic equation. Expand the left side: $x^2 + 2x - 8 = 7$. Subtract 7 from both sides: $x^2 + 2x - 15 = 0$. Now we can factor this and solve: $(x+5)(x-3)= 0$, so $x = -5$ or $x = 3$.
Let's check our answers. First, let's plug $x = -5$ into the left side of the equation. Note that we would then be taking the log of a negative number, which is not allowed. Thus, $x = -5$ is not a valid solution. Next, let's plug $x = 3$ into the left side of the equation: $\ln(3-2)+\ln(3+4)$, which reduces to: $\ln 1 + \ln 7$. Because $\ln 1 = 0$, this becomes $\ln 7$. So, our second solution works. Therefore, even though we found two values of x, the only solution is $x = 3$.
Be careful! Do not automatically throw out a solution where x is a negative number or zero. It is invalid if you end up taking the *log* of a negative number, not if the number you are plugging in is negative.
Let's do another example to see this.

Example 4: Solve for x: $\ln(4-x)+\ln(x+7)= \ln 10$.
The Product Law says that $\log(AB)= \log A + \log B$, so we can rewrite the left side of the equation as $\ln[(4-x)(x+7)] = \ln 10$. Now, because the two logs are equal, we can ignore the log parts and solve the equation $(4-x)(x+7)= 10$. This is a simple quadratic equation. Expand the left side: $-x^2 - 3x + 28 = 10$. Subtract 10 from both sides: $-x^2 - 3x + 18 = 0$. Now, we can factor this and solve: $-(x+6)(x-3)= 0$, so $x = -6$ or $x = 3$.
Let's check our answers. First, let's plug $x = -6$ into the left side of the equation. We get: $\ln(4-(-6))+\ln((-6)+7)$. This reduces to $\ln 10 + \ln 1$. Because $\ln 1 = 0$, this becomes $\ln 10$. Our solution works. Next, let's plug $x = 3$ into the left side of the equation: $\ln(4-3)+\ln(3+7)$. which reduces to: $\ln 1 + \ln 10$. Because $\ln 1 = 0$, this becomes $\ln 10$. This solution works too. Therefore, the solutions are $x = -6$ and $x = 3$. Now we can see that a negative solution can work. Remember, a solution is invalid if it causes us to take the *log* of a negative number.
Sometimes there will be equations that we need to solve where there are logarithms on one side but not the other. We solve them in the same basic way, with one adjustment. Let's do an example.

Example 5: Solve for x: $\log_2 x + \log_2(x-7) = 3$.

The Product Law says that $\log(AB) = \log A + \log B$, so we can rewrite the left side of the equation as $\log_2[x(x-7)] = 3$. Next, let's combine the terms on the left side: $\log_2(x^2 - 7x) = 3$. Now, remember that the definition of a logarithm is $\log_B x = A$ means that $B^A = x$. We can use this to rewrite the equation as $x^2 - 7x = 2^3$ and since $2^3 = 8$, $x^2 - 7x = 8$. If we subtract 8 from both sides, we get $x^2 - 7x - 8 = 0$, which is a simple quadratic equation. Now we can factor this and solve: $(x-8)(x+1) = 0$, so $x = 8$ or $x = -1$.

Let's check our answers. First, let's plug $x = -1$ into the left side of the equation. The $\log_2 x$ will be the log of a negative number, which is not allowed, so we can throw out that answer. Now let's plug $x = 8$ into the left side of the equation. We get: $\log_2 8 + \log_2(8-7)$. This reduces to $\log_2 8 + \log_2 1$. Because $\log_2 1 = 0$, this becomes $\log_2 8 = 3$. Our solution works. Therefore, the solution is $x = 8$.

As we noted before, this was not much different than the previous problems, except that we did not have a logarithm on both sides of the equation. We had to use the definition of a logarithm to rewrite the equation and get rid of the logarithm, which then made it easy to solve.

How about another example?

Example 6: Solve for x: $\log(64x + 120) - \log(x-6) = 2$.

The Quotient Law says that $\log \dfrac{A}{B} = \log A - \log B$, so we can rewrite the left side of the equation as $\log\left(\dfrac{64x+120}{x-6}\right) = 2$. Next, remember that the definition of a logarithm is $\log_B x = A$ means that $B^A = x$. We can use this to rewrite the equation as $\dfrac{64x+120}{x-6} = 10^2$ and since $10^2 = 100$, $\dfrac{64x+120}{x-6} = 100$. Now we can cross-multiply to get $64x + 120 = 100x - 600$. With a little algebra, we get: $36x = 720$, so $x = 20$.

Let's check our answer. We plug $x = 20$ into the left side of the equation: $\log(64(20) + 120) - \log(20-6)$. This simplifies to $\log 1400 - \log 14$. Now we use the Quotient Law to reduce this to $\log \dfrac{1400}{14} = \log 100 = 2$. Our solution works. Therefore, the solution is $x = 20$.

Now let's look at some applications that use exponentials in their setup and will therefore require logarithms for their solutions. One example of where we use logarithms is in Radioactive Decay. Certain isotopes of elements have unstable nuclei, which give off radiation. In doing so, some of these elements change into either different, more stable isotopes, or into different elements. Scientists often look at the *half life* of such elements. The half life is the period of time it takes for half of the element to change from one substance to another. So if, initially there are 100 grams of an element, at the end of the first half life there are 50 grams.

At the end of the second half life, there are 25 grams (half of 50). At the end of the third half life, there are 12.5 grams (half of 25), and so on. This leads to a simple formula: $A = A_0 \left(\dfrac{1}{2}\right)^{\frac{t}{k}}$, where A_0 is the initial amount of the radioactive material, A is the amount after some period of time, t, and the half life is k.

Let's do an example.

Example 7: A radioactive isotope has a half life of 12 days. If there are initially 64 grams, how long will it take to decay to 2 grams?

Let's plug the information into our half life equation. We get: $2 = 64\left(\dfrac{1}{2}\right)^{\frac{t}{12}}$. If we divide both sides by 64, we get $\dfrac{1}{32} = \left(\dfrac{1}{2}\right)^{\frac{t}{12}}$. Because 32 can be written as 2^5, we can rewrite the equation as $\left(\dfrac{1}{2}\right)^5 = \left(\dfrac{1}{2}\right)^{\frac{t}{12}}$. Now we can set the powers equal to each other and solve: $5 = \dfrac{t}{12}$, so $t = 60$ days.

Notice that we did not need logarithms to solve the problem. This is because both sides could be written in terms of a common base. Suppose instead that we wanted to know how long the isotope decays to 10 grams. Now we would need logarithms.

Example 8: A radioactive isotope has a half life of 12 days. If there are initially 100 grams, how long will it take to decay to 10 grams?

Let's plug the information into our half life equation. We get: $10 = 100\left(\dfrac{1}{2}\right)^{\frac{t}{12}}$. If we divide both sides by 100, we get $\dfrac{1}{10} = \left(\dfrac{1}{2}\right)^{\frac{t}{12}}$. Now we have a problem. We can't write 10 as a power of 2 so we will need to use logarithms. We take the log of both sides and we get: $\log\dfrac{1}{10} = \log\left(\dfrac{1}{2}\right)^{\frac{t}{12}}$. Next, we can use the Power Rule to rewrite this as $\log\dfrac{1}{10} = \dfrac{t}{12}\log\dfrac{1}{2}$. Next, we isolate t: $\dfrac{12\log\dfrac{1}{10}}{\log\dfrac{1}{2}} = t$. Then use our calculator: $t \approx 39.863$ days.

Notice that we set up this problem just the way we did the previous one, but we needed to use logarithms to get the answer. In fact, this is one of the main reasons that logarithms were invented: to solve for the unknown power in an exponential equation.

Another place where we see logarithms is in scaling phenomena. Logarithms are very handy when we are trying to compare measurements that vary by great amounts. For example, earthquakes can vary in intensity from practically undetectable to, literally, earth-shattering. The amount of energy released in an earthquake can vary by trillions of joules. It is impractical to create a graph that would show the difference in such intensities using simple linear measurement. The Richter Scale was created to solve this problem. The Richter Scale uses the power of the intensity of the earthquake, rather than the absolute number. In other words, if one event is a million times more powerful than another one, it would only be a difference of 6 times using a scale of log base 10. This is because one million is 10^6, and logarithms look at the exponent. The Richter Scale is $R = \log\dfrac{a}{T} + B$, where a is the amplitude in micrometers of the vertical ground motion at the station that measures the earthquake, T is the period of the seismic wave in seconds, and B is a constant that accounts for the weakening of the seismic wave with increasing distance from the epicenter of the earthquake. Don't worry. It's not important where the measurement of the earthquake comes from. What we are going to do is learn how to compare two different earthquakes to see the difference in energy.

Let's do an example.

Example 9: How many times more powerful is an earthquake of magnitude $R_1 = 6.5$ than one of magnitude $R_2 = 4.5$?

We want to find the ratio of how intense the earthquakes are, which is $\dfrac{a_1}{a_2}$.

We get $R_1 = \log\dfrac{a_1}{T} + B = 6.5$ and $R_2 = \log\dfrac{a_2}{T} + B = 4.5$. How are we going to compare these?

First, let's find the difference between R_1 and R_2. We get: $R_1 - R_2 = \left(\log\dfrac{a_1}{T} + B\right) - \left(\log\dfrac{a_2}{T} + B\right) = 2$. We can simplify this to: $\log\dfrac{a_1}{T} - \log\dfrac{a_2}{T} = 2$. Next,

we use the Quotient Rule to write the left side as a single log: $\log\left(\dfrac{\dfrac{a_1}{T}}{\dfrac{a_2}{T}}\right) = 2$, which

simplifies to $\log\dfrac{a_1}{a_2} = 2$. Now we use the definition of a log to solve for the ratio of

$\dfrac{a_1}{a_2} : \dfrac{a_1}{a_2} = 10^2 = 100$.

This means that if two earthquakes differ by 2 on the Richter Scale, the bigger one is 100 times more powerful than the smaller one, not 2 times more powerful!

Another scale that uses logarithms is the pH scale. This is used to measure acidity levels of a solution, which is done by looking at the concentration of hydrogen ions in the solution in moles per liter. As with earthquakes, these differences can be quite large, so the pH scale uses logarithms to compare the concentrations, rather than using the absolute numbers themselves. More acidic solutions have higher concentrations of hydrogen ions, and lower pH values. The scale does this by using the negative of the logarithm. The pH is $-\log\left[H^+\right]$ ion concentration. Let's do an example.

Example 10: A particular vinegar has a pH of 3.4 and baking soda has a pH of 8.4. What is the ratio of their hydrogen-ion concentrations?

First, let's figure out what the actual concentrations are.

The pH of vinegar is 3.4, so we can find the concentration of hydrogen ions by: $-\log\left[H^+\right] = 3.4$ Now, we solve for the concentration using the definition of a logarithm: $\left[H^+\right] = 10^{-3.4} \approx 3.98\times10^{-4}$ *moles/liter*.

The pH of baking soda is 8.4, so we can find the concentration of hydrogen ions by: $-\log\left[H^+\right] = 8.4$. Now, we solve for the concentration using the definition of a logarithm: $\left[H^+\right] = 10^{-8.4} \approx 3.98\times10^{-9}$ *moles/liter*.

Now if we take the ratio of the two concentrations we get: $\dfrac{3.98\times10^{-4}}{3.98\times10^{-9}} = 10^5$ *moles/liter*.

Here we again see that value of using a logarithmic scale. It enables us to look at a difference in concentrations of $10^5 = 100,000$ and scale it as a difference of 5.

Another type of problem that we can use logarithms to solve is to look at the way a hot object cools down in a medium. This can be calculated using Newton's Law of Cooling. The temperature T of an object at a time t can be modeled by $T(t) = T_m + \left(T_0 - T_m\right)e^{-kt}$, where T_m is the temperature of the medium, T_0 is the initial temperature of the object, and k is a constant.

Let's do an example.

Example 11: A baked potato at a temperature of 90 Celsius is placed in a refrigerator at a temperature of $4°$ Celsius to cool. Five minutes later, the temperature of the potato is $50°$ Celsius. When will the temperature be $30°$ Celsius?

We have $T_0 = 90$, $T_m = 4$, and $t = 5$. We can use this information to solve for the constant k. Once we have done that, we can use the formula to solve for the final time.

First, We plug into the formula: $50 = 4 + (90 - 4)e^{-5k}$. Now, let's solve for k: We get: $50 = 4 + 86e^{-5k}$. Subtract 4 from both sides: $46 = 86e^{-5k}$.

Divide both sides by 86: $\dfrac{46}{86} = e^{-5k}$ and take the natural log of both sides:

$\ln\dfrac{46}{86} = \ln e^{-5k}$, which we can simplify to $\ln\dfrac{46}{86} = -5k$ (Remember that $\ln e^x = x$).

Now we get k is $-\dfrac{1}{5}\ln\dfrac{46}{86} = k \approx 0.12514$. Now we can plug this back into our equation and solve for t, the time when the potato will be 30° Celsius. We get: $30 = 4 + 86e^{-0.12514t}$.

Subtract 4 from both sides: $26 = 86e^{-0.12514t}$.

Divide both sides by 86: $\dfrac{26}{86} = e^{-0.12514t}$ and take the natural log of both sides:

$\ln\dfrac{26}{86} = \ln e^{-0.12514t}$, which we can simplify to $\ln\dfrac{26}{86} = -0.12514t$. Now we can solve

for t. We get: $-\dfrac{1}{0.12514}\ln\dfrac{26}{86} = t \approx 9.56$ minutes.

Now we have seen a variety of ways that we can use logarithms to solve both algebraic and word problems. Are you ready to practice?

Practice Problem Set #10

1) Solve for x: $\log(x+18) - \log(x+3) = \log 4$.

2) Solve for x: $\ln(3x+4) - \ln(2x-5) = \ln 3$.

3) Solve for x: $\log_5(2x-3) + \log_5(x-2) = \log_5 10$.

4) Solve for x: $\ln(x+3) + \ln(x-4) = \ln 8$.

5) Solve for x: $\log_3(x+6) + \log_3(x-2) = 2$.

6) Solve for x: $\log_6(x) + \log_6(x-5) = 1$.

7) Solve for x: $\log_2(5x+18) - \log_2(x-3) = 3$.

8) Solve for x: $\log_5(17x+5) - \log_5(x-3) = 2$.

9) A radioactive isotope has a half life of 5340 years. If there are initially 500 grams, how long will it take to decay to 100 grams?

10) A radioactive isotope has a half life of 12 seconds. If there are initially 25 grams, how long will it take to decay to 1 gram?

11) How many times more powerful was the 1994 Los Angeles earthquake, which had a magnitude of 6.6 on the Richter Scale, than the 2007 Kent, Britain, earthquake, which had a magnitude of 4.3 on the Richter Scale?

12)   How many times more powerful was the 1985 Mexico City earthquake, which had a magnitude of 8.1 on the Richter Scale, than the 2010 Taiwan earthquake, which had a magnitude of 6.4 on the Richter Scale?

13)   Milk has a pH of approximately 6.6 and Milk of Magnesia has a pH of approximately 10.5. What is the ratio of their hydrogen-ion concentrations?

14)   A peach has a pH of approximately 3.9 and banana has a pH of approximately 5.1. What is the ratio of their hydrogen-ion concentrations?

15)   A cup of tea at a temperature of 94° Celsius is placed in a room at a temperature of 18° Celsius to cool. Ten minutes later the temperature of the tea is 32 Celsius. When will the temperature be 20° Celsius?

16)   A pot is removed from an oven at a temperature of 200° Celsius and is placed in a room at a temperature of 24° Celsius to cool. Thirty minutes later the temperature of the pot is 90° Celsius. When will the temperature be 40° Celsius?

Solutions to Practice Problem Set #10

1)   *Solve for x*: $\log(x+18) - \log(x+3) = \log 4$.

Remember the log laws. The Quotient Law says that $\log \dfrac{A}{B} = \log A - \log B$.

We can use this to rewrite the left side of the equation as $\log \dfrac{x+18}{x+3} = \log 4$.

Now, because the two logs are equal, we can ignore the log parts and solve the

equation $\dfrac{x+18}{x+3} = 4$. Cross-multiply: $x + 18 = 4(x+3)$.

  Then solve: $x + 18 = 4x + 12$, so $x = 2$. Let's check our answer. If we plug

$x = 2$ into the left side of the equation, we get: $\log(2+18) - \log(2+3)$, which

reduces to: $\log 20 - \log 5$. Using the Quotient Law, we get $\log \dfrac{20}{5} = \log 4$. Our

solution works. Therefore, the solution is $x = 2$.

2)   *Solve for x*: $\ln(3x+4) - \ln(2x-5) = \ln 3$.

Remember the log laws. The Quotient Law says that $\log \dfrac{A}{B} = \log A - \log B$. We

can use this to rewrite the left side of the equation as $\ln \dfrac{3x+4}{2x-5} = \ln 3$. Now,

because the two logs are equal, we can ignore the log parts and solve the

equation $\dfrac{3x+4}{2x-5} = 3$. Cross-multiply: $3x + 4 = 3(2x-5)$.

Then solve: $3x+4 = 6x-15$, so $x = \dfrac{19}{3}$. Let's check our answer. If we plug $x = \dfrac{19}{3}$ into the left side of the equation, we get: $\ln\left(3\left(\dfrac{19}{3}\right)+4\right) - \ln\left(2\left(\dfrac{19}{3}\right)-5\right)$, which reduces to: $\ln(23) - \ln\left(\dfrac{23}{3}\right)$. Using the Quotient Law, we get $\ln\left(\dfrac{23}{\frac{23}{3}}\right) = \ln 3$. Our solution works. Therefore, the solution is $x = \dfrac{19}{3}$.

3) *Solve for x*: $\log_5(2x-3) + \log_5(x-2) = \log_5 10$.

The Product Law says that $\log(AB) = \log A + \log B$, so we can rewrite the left side of the equation as $\log_5[(2x-3)(x-2)] = \log_5 10$. Now, because the two logs are equal, we can ignore them and solve the equation $(2x-3)(x-2) = 10$. This is a simple quadratic equation. Expand the left and solve: $2x^2 - 7x+6 = 10$. Subtract 10 from both sides: $2x^2 - 7x-4 = 0$. Now we can factor this and solve: $(2x+1)(x-4) = 0$, so $x = -\dfrac{1}{2}$ or $x = 4$. Let's check our answers. First, let's plug $x = -\dfrac{1}{2}$ into the left side of the equation. Note that we would then be taking the log of a negative number, which is not allowed. Thus, $x = -\dfrac{1}{2}$ is not a valid solution. Next, let's plug $x = 4$ into the left side of the equation: $\log_5(2(4)-3) + \log_5(4-2)$, which reduces to: $\log_5 5 + \log_5 2$. Now we use the Product Law to get $\log_5(5 \cdot 2) = \log_5 10$. Our solution works. Therefore, even though we found two values of x, the only solution is $x = 4$ because the other solution is invalid.

4) *Solve for x*: $\ln(x+3) + \ln(x-4) = \ln 8$.

The Product Law says that $\log(AB) = \log A + \log B$, so we can rewrite the left side of the equation as $\ln[(x+3)(x-4)] = \ln 8$. Now, because the two logs are equal, we can ignore the log parts and solve the equation: $(x+3)(x-4) = 8$. This is a simple quadratic equation. Expand the left side: $x^2 - x - 12 = 8$. Subtract 8 from both sides: $x^2 - x - 20 = 0$. Now we can factor this and solve: $(x-5)(x+4) = 0$, so $x = 5$ or $x = -4$. Let's check our answers. First, let's plug $x = -4$ into the left side of the equation. Note that we would then be taking the log of a negative number, which is not allowed. Thus, $x = -4$ is not a valid solution. Next, let's plug $x = 5$ into the left side of the equation: $\ln(5+3) + \ln(5-4)$, which reduces to: $\ln 8 + \ln 1$. Because $\ln 1 = 0$, this becomes $\ln 8$. Our solution works. Therefore, even though we found two values of x, the only solution is $x = 5$ because the other solution is invalid.

5) Solve for x: $\log_3(x+6) + \log_3(x-2) = 2$.

The Product Law says that $\log(AB) = \log A + \log B$, so we can rewrite the left side of the equation as $\log_3[(x+6)(x-2)] = 2$. Next, let's combine the terms on the left side: $\log_3(x^2 + 4x - 12) = 2$. Now, remember that the definition of a logarithm is $\log_B x = A$ means that $B^A = x$. We can use this to rewrite the equation as $x^2 + 4x - 12 = 3^2 = 9$. If we subtract 9 from both sides, we get $x^2 + 4x - 21 = 0$, which is a simple quadratic equation. Now we can factor this and solve: $(x+7)(x-3) = 0$, so $x = -7$ or $x = 3$. Let's check our answers. First, let's plug $x = -7$ into the left side of the equation. The expression $\log_3(x+6)$ will be the log of a negative number, which is not allowed, so we can throw out that answer. Now let's plug $x = 3$ into the left side of the equation. We get: $\log_3(3+6) + \log_3(3-2)$. This reduces to $\log_3 9 + \log_3 1$. Because $\log_3 1 = 0$, this becomes $\log_3 9 = 2$. Our solution works. Therefore, the solution is $x = 3$.

6) Solve for x: $\log_6(x) + \log_6(x-5) = 1$.

The Product Law says that $\log(AB) = \log A + \log B$, so we can rewrite the left side of the equation as $\log_6[x(x-5)] = 1$. Next, let's combine the terms on the left side: $\log_6(x^2 - 5x) = 1$. Now, remember that the definition of a logarithm is $\log_B x = A$ means that $B^A = x$. We can use this to rewrite the equation as $x^2 - 5x = 6^1 = 6$. If we subtract 6 from both sides, we get $x^2 - 5x - 6 = 0$, which is a simple quadratic equation. Now we can factor this and solve: $(x-6)(x+1) = 0$, so $x = 6$ or $x = -1$. Let's check our answers. First, let's plug $x = -1$ into the left side of the equation. The expression $\log_6 x$ will be the log of a negative number, which is not allowed, so we can throw out that answer. Now let's plug $x = 6$ into the left side of the equation. We get: $\log_6 6 + \log_6(6-5)$. This reduces to $\log_6 6 + \log_6 1$. Because $\log_6 1 = 0$, this becomes $\log_6 6 = 1$. Our solution works. Therefore, the solution is $x = 6$.

7) Solve for x: $\log_2(5x+18) - \log_2(x-3) = 3$.

The Quotient Law says that $\log\dfrac{A}{B} = \log A - \log B$, so we can rewrite the left side of the equation as $\log_2\left(\dfrac{5x+18}{x-3}\right) = 3$. Next, remember that the definition of a logarithm is $\log_B x = A$ means that $B^A = x$. We can use this to rewrite the equation as $\dfrac{5x+18}{x-3} = 2^3 = 8$. Now we can cross-multiply to get $5x + 18 = 8x - 24$. With a little algebra, we get: $3x = 42$, so $x = 14$. Let's check our answer. We plug $x = 14$ into the left side of the equation: $\log_2[5(14)+18] - \log_2(14-3)$. This simplifies to $\log_2 88 - \log_2 11$. Now we use the Quotient Law to reduce this to $\log_2\dfrac{88}{11} = \log_2 8 = 3$. Our solution works. Therefore, the solution is $x = 14$.

8) Solve for x: $\log_5(17x+5) - \log_5(x-3) = 2$.

The Quotient Law says that $\log\dfrac{A}{B} = \log A - \log B$, so we can rewrite the left side of the equation as $\log_5\left(\dfrac{17x+5}{x-3}\right) = 2$. Next, remember that the definition of a logarithm is $\log_B x = A$ means that $B^A = x$. We can use this to rewrite the equation as $\dfrac{17x+5}{x-3} = 5^2 = 25$. Now we can cross-multiply to get $17x+5 = 25x - 75$. With a little algebra, we get: $8x = 80$, so $x = 10$. Let's check our answer. We plug $x = 10$ into the left side of the equation: $\log_5(17(10)+5) - \log_5(10-3)$. This simplifies to $\log_5 175 - \log_5 7$. Now we use the Quotient Law to reduce this to $\log_5\dfrac{175}{7} = \log_5 25 = 2$. Our solution works. Therefore, the solution is $x = 10$.

9) *A radioactive isotope has a half life of 5340 years. If there are initially 500 grams, how long will it take to decay to 100 grams?*

Let's plug the information into our half life equation. We get: $100 = 500\left(\dfrac{1}{2}\right)^{\frac{t}{5340}}$. If we divide both sides by 500, we get $\dfrac{1}{5} = \left(\dfrac{1}{2}\right)^{\frac{t}{5340}}$. We can't write 5 as a power of 2 so we will need to use logarithms. We take the log of both sides and we get: $\log\dfrac{1}{5} = \log\left(\dfrac{1}{2}\right)^{\frac{t}{5340}}$. Next, we can use the Power Rule to rewrite this as $\log\dfrac{1}{5} = \dfrac{t}{5340}\log\left(\dfrac{1}{2}\right)$. Next, we isolate t: $\dfrac{5340\log\dfrac{1}{5}}{\log\dfrac{1}{2}} = t$. Then use our calculator: $t \approx 12{,}399$ years.

10) *A radioactive isotope has a half life of 12 seconds. If there are initially 25 grams, how long will it take to decay to 1 gram?*

Let's plug the information into our half life equation. We get: $1 = 25\left(\dfrac{1}{2}\right)^{\frac{t}{12}}$. If we divide both sides by 25, we get $\dfrac{1}{25} = \left(\dfrac{1}{2}\right)^{\frac{t}{12}}$. We can't write 25 as a power of 2, so we will need to use logarithms. We take the log of both sides and we get: $\log\dfrac{1}{25} = \log\left(\dfrac{1}{2}\right)^{\frac{t}{12}}$. Next, we can use the Power Rule to rewrite this as $\log\dfrac{1}{25} = \dfrac{t}{12}\log\left(\dfrac{1}{2}\right)$. Next, we isolate t: $\dfrac{12\log\dfrac{1}{25}}{\log\dfrac{1}{2}} = t$. Then use our calculator: $t \approx 55.7$ seconds.

11) *How many times more powerful was the 1994 Los Angeles earthquake, which had a magnitude of $R_1 = 6.6$ on the Richter Scale, than the 2007 Kent, Britain, earthquake, which had a magnitude of $R_2 = 4.3$ on the Richter Scale?*

We want to find the ratio of how intense the earthquakes are, which is $\dfrac{a_1}{a_2}$.

We get $R_1 = \log\dfrac{a_1}{T} + B = 6.6$ and $R_2 = \log\dfrac{a_2}{T} + B = 4.3$. How are we going to compare these?

First, let's find the difference between R_1 and R_2. We get: $R_1 - R_2 = \log\dfrac{a_1}{T} + B - \log\dfrac{a_2}{T} + B = 2.3$. We can simplify this to: $\log\dfrac{a_1}{T} - \log\dfrac{a_2}{T} = 2.3$.

Next, we use the Quotient Rule to write the left side as a single log: $\log\dfrac{\frac{a_1}{T}}{\frac{a_2}{T}} = 2.3$, which simplifies to $\log\dfrac{a_1}{a_2} = 2.3$. Now we use the definition of a log to solve for the ratio of $\dfrac{a_1}{a_2}$: $\dfrac{a_1}{a_2} = 10^{2.3} \approx 200$.

12) *How many times more powerful was the 1985 Mexico City earthquake, which had a magnitude of 8.1 on the Richter Scale, than the 2010 Taiwan earthquake, which had a magnitude of 6.4 on the Richter Scale?*

We want to find the ratio of how intense the earthquakes are, which is $\dfrac{a_1}{a_2}$.

We get $R_1 = \log\dfrac{a_1}{T} + B = 8.1$ and $R_2 = \log\dfrac{a_2}{T} + B = 6.4$. How are we going to compare these?

First, let's find the difference between R_1 and R_2. We get: $R_1 - R_2 = \log\dfrac{a_1}{T} + B - \log\dfrac{a_2}{T} + B = 1.7$. We can simplify this to: $\log\dfrac{a_1}{T} - \log\dfrac{a_2}{T} = 1.7$.

Next, we use the Quotient Rule to write the left side as a single log: $\log\dfrac{\frac{a_1}{T}}{\frac{a_2}{T}} = 1.7$, which simplifies to $\log\dfrac{a_1}{a_2} = 1.7$. Now we use the definition of a log to solve for the ratio of $\dfrac{a_1}{a_2}$: $\dfrac{a_1}{a_2} = 10^{1.7} \approx 50$.

13) *Milk has a pH of approximately 6.6 and Milk of Magnesia has a pH of approximately 10.5. What is the ratio of their hydrogen-ion concentrations?*

First, let's figure out what the actual concentrations are.

The pH of milk is 6.6, so we can find the concentration of hydrogen ions by: $-\log\left[H^+\right] = 6.6$. Now, we solve for the concentration using the definition of a logarithm: $\left[H^+\right] = 10^{-6.6} \approx 2.51 \times 10^{-7}$ *moles/liter.*

The pH of Milk of Magnesia is 10.5, so we can find the concentration of hydrogen ions by: $-\log\left[H^+\right]=10.5$. Now, we solve for the concentration using the definition of a logarithm: $\left[H^+\right]=10^{-10.5}\approx3.16\times10^{-11}$ *moles/liter*. Now if we take the ratio of the two concentrations we get: $\dfrac{2.51\times10^{-7}}{3.16\times10^{-11}}\approx8000$ *moles/liter*.

14) *A peach has a pH of approximately 3.9 and a banana has a pH of approximately 5.1. What is the ratio of their hydrogen-ion concentrations?*

First, let's figure out what the actual concentrations are.

The pH of a peach is 3.9, so we can find the concentration of hydrogen ions by: $-\log\left[H^+\right]=3.9$. Now, we solve for the concentration using the definition of a logarithm: $\left[H^+\right]=10^{-3.9}\approx1.26\times10^{-4}$ *moles/liter*.

The pH of a banana is 5.1, so we can find the concentration of hydrogen ions by: $-\log\left[H^+\right]=5.1$. Now, we solve for the concentration using the definition of a logarithm: $\left[H^+\right]=10^{-5.1}\approx7.94\times10^{-6}$ *moles/liter*.

Now if we take the ratio of the two concentrations we get: $\dfrac{1.26\times10^{-4}}{7.94\times10^{-6}}=$ 16 *moles/liter*.

15) *A cup of tea at a temperature of 94° Celsius is placed in a room at a temperature of 18° Celsius to cool. Ten minutes later the temperature of the tea is 32° Celsius. When will the temperature be 20° Celsius?*

This can be calculated using Newton's Law of Cooling. The temperature T of an object at a time t can be modeled by $T(t)=T_m+\left(T_0-T_m\right)e^{-kt}$, where T_m is the temperature of the medium, T_0 is the initial temperature of the object, and k is a constant. Here, we have $T_0=94$, $T_m=18$, and $t=10$. We can use this information to solve for the constant k. Once we have done that, we can use the formula to solve for the final time.

First, We plug into the formula: $32=18+(94-18)e^{-10k}$.

Now, let's solve for k: We get: $32=18+76e^{-10k}$. Subtract 18 from both sides: $14=76e^{-10k}$.

Divide both sides by 76: $\dfrac{14}{76}=e^{-10k}$, and take the natural log of both sides: $\ln\dfrac{14}{76}=\ln e^{-10k}$, which we can simplify to $\ln\dfrac{14}{76}=-10k$ (Remember that $\ln e^x=x$). Now we get k is $-\dfrac{1}{10}\ln\dfrac{14}{76}=k\approx0.16917$. We can then plug this back into our equation and solve for t, the time when the tea will be 20° Celsius.

We get: $20=18+76e^{-0.16917t}$.

Subtract 18 from both sides: $2=76e^{-0.16917t}$.

Divide both sides by 76: $\dfrac{2}{76} = e^{-0.16917t}$, and take the natural log of both sides: $\ln\dfrac{2}{76} = \ln e^{-0.16917t}$, which we can simplify to $\ln\dfrac{2}{76} = -0.16917t$. Now we can solve for t. We get: $-\dfrac{1}{0.16917}\ln\dfrac{2}{76} = t \approx 21.5$ minutes.

16) *A pot is removed from an oven at a temperature of 200° Celsius and is placed in a room at a temperature of 24° Celsius to cool. Thirty minutes later the temperature of the pot is 90° Celsius. When will the temperature be 40° Celsius?*

This can be calculated using Newton's Law of Cooling. The temperature T of an object at a time t can be modeled by $T(t) = T_m + (T_0 - T_m)e^{-kt}$ where T_m is the temperature of the medium, T_0 is the initial temperature of the object, and k is a constant. Here, we have $T_0 = 200$, $T_m = 24$, and $t = 30$. We can use this information to solve for the constant k. Once we have done that, we can use the formula to solve for the final time.

First, We plug into the formula: $90 = 24 + (200 - 24)e^{-30k}$.

Now, let's solve for k. We get: $90 = 24 + 176e^{-30k}$. Subtract 24 from both sides: $66 = 176e^{-30k}$.

Divide both sides by 176: $\dfrac{66}{176} = e^{-30k}$, and take the natural log of both sides: $\ln\dfrac{66}{176} = \ln e^{-30k}$, which we can simplify to $\ln\dfrac{66}{176} = -30k$ (Remember that $\ln e^x = x$). Now we get k is $-\dfrac{1}{30}\ln\dfrac{66}{176} = k \approx 0.0327$. Now we can plug this back into our equation and solve for t, the time when the pot will be 40° Celsius.

We get: $40 = 24 + 176e^{-0.0327t}$.

Subtract 24 from both sides: $16 = 176e^{-0.0327t}$

Divide both sides by 176: $\dfrac{16}{176} = e^{-0.0327t}$ and take the natural log of both sides: $\ln\dfrac{16}{176} = \ln e^{-0.0327t}$, which we can simplify to $\ln\dfrac{16}{176} = -0.0327t$. Now we can solve for t. We get: $-\dfrac{1}{0.0327}\ln\dfrac{16}{176} = t \approx 73.3$ minutes.